Managing
the Energy
Transition

Managing the Energy Transition:

A System Dynamics Search for Alternatives to Oil and Gas

Roger F. Naill
Dartmouth System Dynamics Group
Thayer School of Engineering
Dartmouth College

Ballinger Publishing Company ● Cambridge, Massachusetts
A Subsidiary of J.B. Lippincott Company

This book is printed on recycled paper.

International Standard Book Number: 0-88410-608-X

Library of Congress Catalog Card Number: 76-52752

Printed in the United States of America

Library of Congress Cataloging in Publication Data

Naill, Roger F
  Managing the energy transition.
  Originally presented as the author's thesis, Dartmouth.
  Bibliography: v. 1, p.
  Includes index.
  1. Power resources—United States—Mathematical models. 2. Energy consumption—United States—Mathematical models. 3. Energy policy—United States—Mathematical models. I. Title.

TJ163.25.U6N33   333.7   76-52752
ISBN 0-88410-608-X

To my family:
Carol, Sara, and Megan

Contents

List of Figures

Foreword

The nation exhibits a growing sense of concern and futility as energy legislation designed to lessen dependence on imports is passed, while at the same time oil imports continue to rise. The rapid deterioration of the United States energy balance, and the resulting vulnerability of our economic system to foreign influence, has focused national concern on United States energy policy. In response, the federal government has invested hundreds of millions of dollars to develop the proper tools to assess the situation and design and implement effective policy. Energy modeling has become big business.

However, most of the models currently in use focus so narrowly on specific issues that they cannot measure the comprehensive interactions that make the United States energy system so difficult to manage. The system itself is characterized by a number of perplexing properties that complicate standard economic or geological analyses.

Inertia. The energy system is stubbornly insensitive to policy change. Even when a substantial change is implemented, behavior often proceeds along the former course for long periods of time. This "inertia" is caused primarily by the long delays inherent in shifting capital investment in the energy system.

Interconnectedness. Energy policy-making is currently highly fragmented, with separate legislative committees and subcommittees responsible for different branches of the energy system. Yet the system itself is highly interconnected. For example, any legislation directed toward Western coal mining will also influence in some way all the other elements of the system, including nuclear power growth, oil imports, and total energy demand.

Side Effects. Energy policy changes designed to improve some aspects of the system often create unforeseen side effects. The energy legislator must continually juggle difficult tradeoffs, only some of which are obvious, in considering any policy change.

Short-term versus long-term tradeoffs. Energy policies often produce long-term responses directly opposite to their more immediate short-term effects. This means that political leaders faced with the task of legislating effective long-term energy policies sometimes have to explain a "worse before better" situation to their constituents. Voter pressure and political expediency tend to favor short-term solutions, setting the stage for likely long-term degeneration of the United States energy system.

Inertia, interconnectedness, side effects, and short-term/long-term tradeoffs are all characteristics of a class of systems called *multiple-loop nonlinear feedback systems.* Dr. Naill's analysis employs a modeling paradigm called *system dynamics*, developed to examine the behavior of such complex systems. System dynamics was pioneered by Professor Jay W. Forrester at M.I.T. in 1956, originally as a methodology applicable to problems of industrial management. However, it has since proved effective in analyzing social systems as diverse as population control mechanisms in primitive tribes and global food supply.

System dynamics integrates three distinct fields of endeavor to analyze social systems: feedback control theory, organizational behavior, and computer technology. The focus on the interrelationship between the state of the system and its rate of change through time (implying a feedback loop) is drawn from feedback control theory. The idea that social systems—systems involving human decision-making processes—could be modeled with the same techniques as physical systems stems from the field of organizational behavior. Advanced computer technology allows the manipulation of complicated data inputs within the model structure in order to project system behavior under varying conditions.

Thus the problem—the United States energy system—and the tool—system dynamics modeling—are ideally suited for each other. Because Dr. Naill's model, COAL2, does a good job of reproducing the historical behavior of the system, there is reason to believe the validity of its assessments of present and future energy dynamics.

Dennis Meadows, co-author of *Limits to Growth* and Associate Professor at Dartmouth College's Thayer School of Engineering, conducted in 1976 a one-week exhaustive review of national energy models that have thus far been developed. His assessment of COAL2 was that it "emerged as the most comprehensible, well-balanced, and tractable of the group. It is easily updated and its underlying assumptions are clear. The ease with which it can be analyzed and its low operation and maintenance costs make it the ideal basis for more detailed and short-term studies of interest to specific organizations. I believe COAL2 can remain a useful basis for decision-making over many years."

The COAL2 model has already gained national recognition. It has been the subject of a cover article in one of the nation's leading technical journals, the starting point for testimony before several Congressional subcommittees, and the basis for reports to the FEA, ERDA, the GAO, and other federal agencies. The program for the model has been translated into FORTRAN and is now widely distributed for use in classroom teaching. This volume will provide easy access to the complete COAL2 model—its assumptions, structure, and policy implications—for an even wider and increasingly non-technical audience.

John D. Moody
Petroleum Consultant

Preface

This book presents a system dynamics model of energy supply and demand, COAL2, to aid in the formulation of United States energy policy. The model fucuses on such long-term factors as energy-demand growth, resource depletion, price effects, lead times, and financial and environmental constraints to the development of new energy sources. Using the COAL2 model, we have concluded that the United States energy problem is much more persistent than current thinking acknowledges, and much less amenable to straightforward solutions. This volume is intended to increase public understanding of the underlying causes of the United States energy imbalance and the effectiveness of various policies designed to improve the behavior of the energy system.

The seed for this book was planted by the Club of Rome in 1970. The Club, a group of scientists and industrialists concerned about the future of the globe, initiated at M.I.T. a world modeling effort directed by Professor Dennis L. Meadows and entitled "Project on the Predicament of Mankind." An international group of ten scientists and students constructed a computer simulation model, WORLD3, to understand the long-term interactions which cause and limit exponential growth in population and industrial output. The nontechnical report of that project, *Limits to Growth*, concluded that the world system is rapidly approaching a number of physical and social limits which, singly or in combination, could cause collapse over the long term.

A number of research projects focusing on specific growth-related problems followed from the Club of Rome project. One of these was my study of United States natural gas supply, included in a second volume entitled *Toward Global Equilibrium: Collected Papers*, edited by Dennis and Donella Meadows. This study's conclusions are painfully well-accepted today: namely, that domestic gas production, having peaked in 1973, will decline well below current production levels by the year 2000. The projected rapid decrease in availability of natural gas led our group to ask: will United States economic growth be impeded by an "energy limit" similar to those suggested in *Limits to Growth?*

To answer this question, the Dartmouth System Dynamics Group, an off-shoot of the original M.I.T. group, received in 1972 a three-year contract from the National Science Foundation to study what we came to describe as the United States "energy transition" problem. COAL1 and COAL2 are the models we built for that project, and this book summarizes our major insights and conclusions.

In 1975 ERDA provided us with additional support to construct a new model, FOSSIL1, for use in government energy planning. This model, used in conjunction with other energy analyses and good common sense, will help analyze and design new energy legislation to enable the United States to weather the "energy transition" smoothly.

This work was originally written in partial fulfillment of the requirements for the degree of Doctor of Philosophy at the Thayer School of Engineering at Dartmouth College. Professor Dennis L. Meadows, who served as chairman of my dissertation committee, has been a continuous source of guidance and support over the past few years. I am further indebted to the other members of my thesis committee—Professor Alvin O. Converse, Dean Carl F. Long, and John Moody—for their advice regarding the content and presentation of the material.

Several members of the Dartmouth System Dynamics Group also merit special thanks. Dr. Robert E. Sweeney, Jerome B. Doolittle, and Marion McCollom edited the manuscript and made numerous suggestions to improve the structure and logic of the presentation. Substudies by Leif Ervik, Frederick A. Ford, Jay Jacobsen, Michael Maddox, and Joel Rahn helped provide the conceptual foundation for the COAL1 and COAL2 models. Steve Flanders and Heidi Deering prepared the diagrams for the volume. Jean Graf, Lynn Eberhardt, and Barbara Ferraro typed and retyped each new draft of the manuscript with exemplary patience.

Roger F. Naill

A Note to the Reader

Most people are accustomed to measuring energy usage in terms of the physical quantities of the energy form used: tons of coal, barrels of oil, cubic feet of natural gas, or kilowatt-hours of electricity. To discuss the overall energy system, as this book does, it is helpful to convert these various units of measurement to a common unit of convenient size. One property that all fuels have in common is their *heat value:* in the English system of measurement, the basic unit is the Btu. A Btu is a British thermal unit, defined as the amount of heat needed to raise the temperature of one pound of water one degree Fahrenheit.

The following table gives some useful conversion factors:

oil:	5,800,000 Btu/barrel
natural gas:	1,030 Btu/standard cubic foot (SCF)
coal:	22,500,000 Btu/ton[a]
electricity:	3,412 Btu/kilowatt-hour at 100 percent conversion efficiency
	10,500 Btu/kilowatt-hour at 32.5 percent conversion efficiency

Note that because of the inherent conversion losses in generating electricity from conventional heat sources, about 3 to 3.3 Btu's of energy from coal, oil, gas, or uranium are required to produce one Btu of electrical output.

---

[a]For medium heating value coal at 11,250 Btu/lb.

Since the Btu is a small unit compared to the amounts of energy currently used by the United States each year, it is more convenient to use the *quad*—one quadrillion Btu or 10^{15} Btu—when discussing the energy system.

**Managing
the Energy
Transition**

 *Chapter 1*

The Energy Transition Problem

The fundamental fact remains that the United States has entered a new age of energy, and we have not yet adjusted our habits, expectations, and national policies to the new age.

Ford Energy Policy Project
A Time to Choose, p. 1

OVERVIEW

In 1956 M. King Hubbert predicted that United States crude oil and natural gas production would peak in 1970 (Hubbert 1956, pp. 17-18). His prediction, based on geological estimates of the nation's oil and gas resources, has proven remarkably accurate. Production of oil and gas in 1975 was about 11 percent below the 1972 peak, and is likely to drop rapidly during the 1980s (FEA 1976, pp. 55, 119). During the next few decades, the United States must undergo a transition from dependence on domestic oil and gas to reliance on new sources of energy.

Twice previously in its history the United States has restructured its economy to move from primary reliance on one fuel to another. Figure 1-1 shows that the transitions from wood to coal, and from coal to oil and gas, have each taken over 60 years to complete. Because each transition introduced a more efficient, convenient form of energy, past transitions have encouraged growth in the other sectors of the United States economy. But the current transition is different—there is no cheap, abundant, and environmentally acceptable alternative to the use of domestic oil and gas. For lack of a domestic alternative, the United States is currently following a path that has led to excessively high imports during the early transition period, and will lead to even greater imports as the transition progresses. To counter this trend toward imports, the federal government is currently considering the following proposals:

- mandatory conservation measures
- oil import quotas or tariffs
- decontrol of oil and gas prices
- accelerated development of nuclear power
- lower air quality standards
- accelerated synthetic fuels research and development
- federal loan guarantees or price supports for synthetic fuels commercialization
- rate reform for electric utilities
- federal subsidies for coal development

In each case, government policymakers must determine whether federal intervention is warranted: benefits from lower imports must be weighed against the economic and environmental costs of each program. To analyze the long-term effects of federal policy on the United States energy system, a system dynamics[a] model, entitled COAL2, has been constructed at Dartmouth College's Thayer School of Engineering. This model emphasizes the complex *causal* mecha-

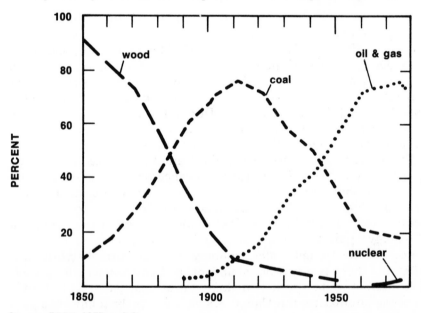

Source: ERDA 1975, p. S-2.

Figure 1-1. Major energy sources, 1850-1974

[a]For an explanation of system dynamics, see J.W. Forrester, *Industrial Dynamics*, Cambridge, Mass.: M.I.T. Press, 1961; *Principles of Systems*, Cambridge, Mass.: Wright-Allen Press, 1968; or Dennis L. Meadows, William W. Behrens III, Donella H. Meadows, Roger F. Naill, Jørgen Randers, Erich K. Zahn, *Dynamics of Growth in a Finite World*, Cambridge, Mass.: Wright-Allen Press, 1974.

nisms that determine the production and consumption of energy over the long term. In this book, the model is used to assess the relative effectiveness of the alternative energy strategies currently under consideration by United States policymakers.

THE TRANSITION PROBLEM

United States energy consumption has grown at an average of 3 percent annually for the past 70 years, and at 3.5 percent over the past 25 years. As shown in Figure 1-2, this extraordinary period of sustained growth has been accomplished almost entirely through increased consumption of petroleum and natural gas. Oil and gas usage grew from negligible amounts in 1900 to over 75 percent of gross United States energy inputs in 1975. This inexpensive, readily-available supply of energy stimulated a high rate of growth in energy consumption, accompanied by a rapid rise in the material standard of living. But domestic production of oil and gas has been declining since 1972, and there are no prospects for a significant upturn in production rates. The recent reappraisal of the nation's oil and gas resources by the United States Geological Survey and the National Academy of Sciences (USGS 1975; COMRATE 1975a) led the Energy Research and Development Administration (ERDA) to conclude that current production rates will be "difficult to maintain" despite the expected contribution from offshore and Alaskan deposits (ERDA 1975, p. S-2).

Ultimate energy sources such as nuclear fusion, solar, wind, ocean thermal gradient, bioconversion, and geothermal are the most desirable alternatives to oil and gas in the far future. But they probably will be unable to provide more than 10 to 20 percent of the nation's energy demand by the year 2000 (ERDA 1975, p. B-8). The vast social and economic changes implied by a nationwide transition to ultimate sources may not permit them to carry the bulk of the country's energy burden before 2050.

The extent of the problem posed by the expected decrease in domestic oil and gas production, combined with the slow increase in production from ultimate sources, depends heavily on future energy demand. Clearly any reduction in the historical rate of demand growth would help alleviate the United States energy problem. Yet even the "Zero Energy Growth" scenario of the Ford Foundation Energy Policy Project assumed that consumption would grow to levels one-third above those of 1975 before it stabilized (Ford 1974a, p. 98). The 1975 ERDA National Energy Plan projects at least a 60 percent increase in energy consumption by the year 2000 (ERDA 1975, p. B-8).

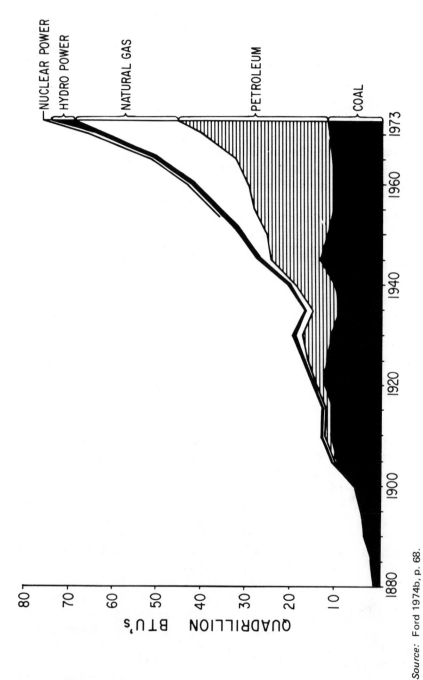

Figure 1-2. Energy consumption, 1880-1973

Source: Ford 1974b, p. 68.

Continued demand growth, depletion of oil and gas resources, and long delays in the implementation of ultimate energy sources raise the possibility of a significant "energy gap" between energy demand and domestic energy production, as shown in Figure 1-3. Under these conditions, the United States must resort to massive energy imports to balance supply and demand during the transition period. The consequences of heavy dependence on imports could well be several decades of rising prices, increasing government intervention in both supply and demand decisions, supply interruptions, and stagnation or decline in the material standard of living. While certainly not inevitable, such an unsatisfactory transition phase is entirely plausible, and, according to COAL2, is the most likely outcome of current policies and trends.

THE COAL2 MODEL

To reduce the energy gap of Figure 1-3 to tolerable levels, the United States needs an intensive effort that focuses national energy policy

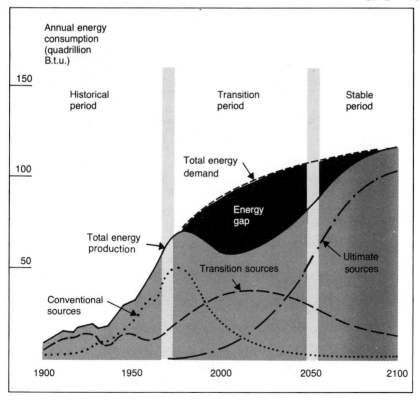

Figure 1-3. The United States energy transition problem

on the transition problem. This effort requires planning well in advance of the transition period because of the long time required for policies to have their full effect. To design an effective energy transition strategy, some fundamental questions must first be answered:

- Is energy independence feasible, and if so, by when?
- Should a national energy policy emphasize conservation or increased supply?
- Which transition energy source should be accelerated?

The purpose of the COAL2 model is to assess the potential magnitude of the United States energy transition problem, and to identify policies that may lead to a smooth transition.

We conclude that in the short term (to 1985), the United States energy problem cannot be solved. No set of policy changes can avoid a significant increase in dependence on oil imports through 1985, due to the momentum of past energy policies and the inherent delays before new policies become effective. Further, neither demand nor supply policies alone will ameliorate the transition problem sufficiently. The transition will require policies that both stabilize demand and accelerate supply to achieve an acceptable transition. And finally, the most promising transition fuel is coal. With emphasis on coal and a significant reduction of demand, the long-term energy gap could be closed by the year 2000.

Many well-informed experts and organizations will disagree with the conclusions of this study. The fact that the conclusions are drawn from a formal computer model can improve the quality of the ongoing debate and dialogue about our energy future. Any criticism implies that some assumption in the model could be improved. We acknowledge this possibility, and by publishing the assumptions of the model, encourage such efforts at improvement.[b] The formal model focuses the debate on factual issues, rather than disputes of opinion among scholars and politicians. The result of such a process will, we trust, be an improved U.S. energy policy.

STRUCTURE OF THE BOOK

The following chapters present, in nontechnical language, the assumptions and behavior of the COAL2 model. Chapter Two provides a brief overview of the model, and describes the policy "levers" built into it. Chapters Three through Six present the four sectors of

---

[b]In fact, the COAL2 model is already in the process of improvement by the Dartmouth System Dynamics Group under contract from the Energy Research and Development Administration.

COAL2: demand, oil and gas, electricity, and coal. Each sector describes, in effect, a submodel, and can be read independently of the complete model. Readers who are interested in specific policies are referred to the simulation runs in these chapters.

In Chapter Seven, the complete COAL2 model is used to design an improved long-term U.S. energy strategy. Technological, economic, social, and environmental policies are examined to find the best way to close the domestic energy gap. Chapter Eight summarizes the policy implications of the COAL2 model.

At the end of each sector chapter is a list of the DYNAMO equations needed to reproduce the runs in that chapter. Appendix B lists equations for the whole model.[c] (These equations are sufficient documentation to reproduce the runs in the book.) Appendix C is a list of definitions of variables used in the model.

---

[c]DYNAMO is available from Pugh-Roberts Associates, 5 Lee Street, Cambridge, Massachusetts, or contact the Dartmouth System Dynamics Group, Thayer School of Engineering, Hanover, New Hampshire, 03755.

Chapter 2

A Brief Description of COAL2

ENERGY DYNAMICS

The flow of energy from primary resources to satisfaction of end-use demand represented in the COAL2 model is illustrated in Figure 2-1. From this perspective, the U.S. energy system converts and processes primary energy into three products for end use: electricity, coal, and a combination of oil and gas. When energy is converted from one form to another (for example, coal to synthetic fuels or fossil and nuclear energy to electricity), considerable energy is lost. Therefore the net energy delivered to the consumer is considerably less than the primary or gross energy extracted. In 1974, 18 percent of United States primary energy consumption was lost in energy conversion processes (USDI 1975, p. 8).

Each of the solid arrows in the energy network of Figure 2-1 represents *energy flow rates* (measured in Btu's per year in COAL2). Figure 2-2 shows that each energy flow rate can be thought of as controlled by a valve (✗ in the diagram). These valves open and close as time passes. For example, the end-use demand valves have been steadily opening at an average rate of 3 percent per year for the past 25 years, due to increased production of energy-consuming goods and services. Increased end-use demands for energy have necessitated a rise in production and conversion of primary energy resources.

Each of the extraction and conversion energy flows shown in Figure 2-2 represents a production process, controlled by *a produc-*

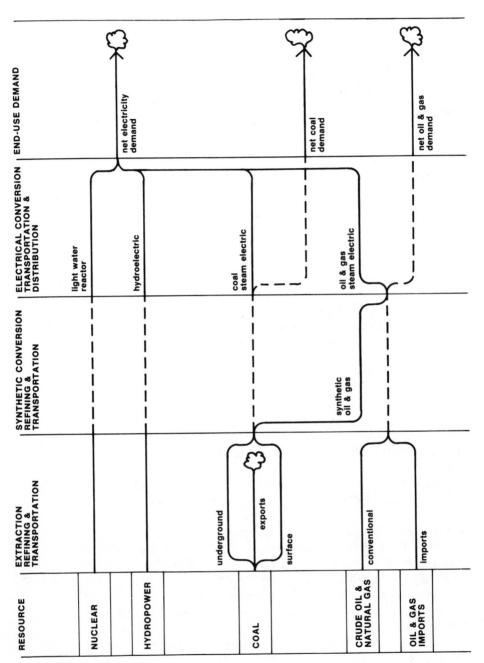

Figure 2-1. United States energy flow network

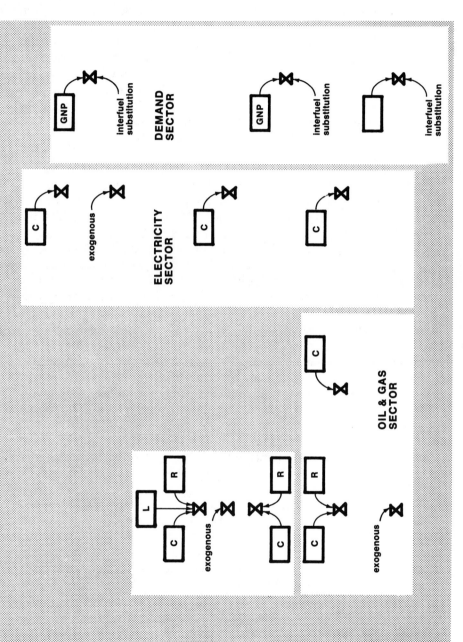

Figure 2-2. Determinants of energy flows in COAL2

tion function in the COAL2 model. (For a discussion of production functions, see Ervik 1974.) The COAL2 production functions relate resource extraction outputs (coal, oil and gas), synthetic fuel conversion outputs (synthetic oil and gas), and electrical conversion outputs (electricity), to specific labor, capital, or resource inputs (represented by ⌊L⌋ , ⌊C⌋ , and ⌊R⌋ in Figure 2-2). These inputs are formulated as levels (or state variables), and are the basic building blocks of the COAL2 model. Figure 2-2 indicates the inputs (or levels) that determine each energy flow rate (⋈) in the COAL2 model.

It is easy to visualize a "snapshot" of the system shown in Figure 2-2 at any point in time: energy capital, labor, and resources are constant, allowing a fixed energy flow rate through each of the valves. (Such snapshots are provided by, for example, the Brookhaven Reference Energy System; see Cherniavsky, 1974.) Yet the availability of energy resources, capital, and labor can change in complex ways over time: oil and gas resources are depleted, energy investment decisions and capital availability can shift rapidly, or underground coal labor may become a scarce input. The structure of the COAL2 model reproduces the complex interaction of geologic, economic, environmental, and technological factors that control the dynamics of energy labor, capital, and resources.

Ideally, domestic energy flows from primary resources to end-use demand should be balanced: there would be little need for imports as an energy input. Yet oil and gas depletion and regulation of domestic oil and gas prices have reduced U.S. energy supplies well below demand over the past 25 years. Oil imports have increased to satisfy almost 20 percent of U.S. energy consumption in 1975 (FEA 1976, p. xxiii). Energy policymakers need to understand both the cause for such a rapid increase in energy imports and the effect of alternative energy policies on the future behavior of imports, energy prices, production, and demand.

Figure 2-2 implies that the influence of federal and corporate policymakers over the flow of energy through our energy system is indirect, at best. Energy policies (such as price deregulation, government-sponsored research and development, or environmental legislation) ordinarily do not have a direct influence on the flow of energy. Over the long term they can only affect the availability of the factors of production (the levels shown in Figure 2-2). Even then they must operate within the economic, social, and institutional decision-making structure that constrains the behavior of the United States energy system. The COAL2 model simulates this structure, allowing the model user to alter the model's decision rules to match proposed new energy policies.

COAL2 MODEL STRUCTURE

The COAL2 model is divided into four major sectors:

Energy demand—including the effects of GNP and energy prices on demand growth and the shift in demand among fuels

Oil and gas—incorporating depletable oil and gas supplies, imports, and synthetics derived from coal

Electricity—including generation of electricity from oil and gas-fired utilities, coal-fired utilities, nuclear, and hydroelectric power plants

Coal—including surface-mined and underground coal production

The sectors of the United States energy system are highly interdependent: any change in energy policy in one sector will affect the future behavior of all the energy flows shown in Figure 2-2. Figure 2-3 illustrates the major interactions included in the COAL2 model.

The *Demand Sector* (depicted at the top of Figure 2-3) determines the behavior of net energy demand and the *form* of energy demanded (oil and gas, electricity, or coal). Net energy demand is defined as energy delivered for use by the energy-consuming sectors of the economy—industrial, transportation, household and commercial—and consists of direct fuel use plus purchased electricity. Net energy demand grows as GNP increases, but its growth slows as energy prices rise. The relative contributions of oil and gas, electricity, and coal that satisfy net energy demand can shift over time as a function of relative prices and fuel convenience. For instance, coal has lost its large share of the United States market due to its inconvenience, while electricity has increased its share of the market due both to its convenience and its decline in price relative to other fuels.

The *Oil and Gas Sector* must satisfy demands for oil and gas from both final consumers and electric utilities, where oil and gas are burned to generate electricity. United States demand for oil and gas can be satisfied domestically from two sources: conventional oil and gas wells or conversion of coal to synthetic fuels. The oil and gas sector includes a financial subsector that generates new capital investments, and two production subsectors that allocate investment to conventional or synthetic production facilities. Both the economic effects of oil and gas resource depletion and the technological delays in the development of synthetic fuels are included in the model structure.

The *Electricity Sector* simulates both the financing and fuel-mix decisions made by utilities over the next 35 years. Utility financing is

heavily influenced by regulation, which is modeled explicitly in
COAL2. Because hydropower development has nearly reached saturation in the United States (FEA 1974, p. 128), investors must choose
among nuclear power, coal-fired plants, or oil and gas-fired plants to
provide most of the planned additions to electric generation capacity. The fuel-mix decision is influenced by capital costs, fuel costs,
and environmental considerations in the COAL2 model.

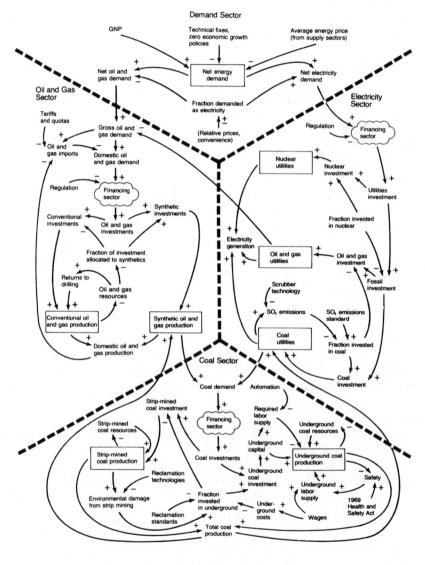

Figure 2-3.   Basic structure of COAL2 model

The *Coal Sector* must satisfy future coal demand from direct burning of coal, coal-fired electric utilities, or synthetic conversion facilities. Coal is produced from two sources in Figure 2-3: surface mines and underground mines. The COAL2 model differentiates between surface and underground coal production because the resource, capital, labor, and environmental characteristics of each production process are significantly different.

As the structure of COAL2 shown in Figure 2-3 illustrates, the many constraints on energy demand and supply are parts of an integrated system. Any policy changes—affecting energy demand, prices, synthetic fuel availability, power plant emission standards, strip-mining legislation, or health and safety standards—will ultimately affect the evolution of the entire energy system. COAL2 can keep track of all these complex interactions, assessing the effects of energy policies on the total energy system.

COAL2 AND ENERGY POLICY

The dominant issue in the current energy debate is how to shrink oil imports to a level that reduces the country's vulnerability to a new oil embargo. The federal government is currently considering a number of policies designed to reduce imports (listed in Figure 2-4). In each case, government policymakers must determine whether federal intervention is warranted. Even though imports might be reduced, the financial and environmental costs of intervention could be so great that the country might be better off to accept high oil imports as a fact of life, with perhaps less freedom of action for the United States in its foreign policy.

To provide a base case that assesses the need for federal intervention in the normal workings of the energy system, a *reference projection* of the COAL2 model is first developed. The reference projection shows the most probable behavior of energy demand and the contribution of oil and gas, synthetic fuels, coal, nuclear power, and imports to United States energy supply from 1950 to the year 2010, *given no major changes in United States or foreign energy policies.*

Subsequent runs in each sector chapter test the effects of proposed policies on the behavior of each sector. Chapter Seven combines all the sectors into one integrated structure, and generates a number of alternative, policy-dependent energy projections. Chapter Eight concludes with recommendations for the combination of policies which we feel offer the best tradeoff among energy independence, economic development, energy price increases, and environmental quality.

**Resource Extraction,
Refining, Transportation**

- Nuclear fuel subsidies
- Oil import quotas, embargoes
- Foreign oil tariffs
- Enhanced oil and gas recovery
- Oil and gas price deregulation
- 1969 Coal Mine Health & Safety Act
- Ban on surface mining
- Surface mining restrictions
  a. Steep slope restrictions
  b. Federal surface coal reclamation standards
  c. Surface-mined coal tax
- Coal investment incentives
  a. Loan guarantees
  b. Coal price support

**Synthetic Conversion,
Refining, Transportation**

- Accelerated synthetic fuels RD&D
- Accelerated synthetic fuels commercialization incentives (price supports or loan guarantees)

**Electricity Conversion,
Transportation, Distribution**

- Utility rate relief
- Improved utility load management
- Accelerated nuclear program
- Relaxation of SO_2 emission standards
- Accelerated implementation of SO_2 emission reduction technologies
- Nuclear moratorium

End-Use Demand

- Accelerated conservation policies
- Zero Energy Growth

Figure 2-4.   COAL2 policy options

The Demand Sector

PURPOSE OF THE DEMAND SECTOR

The energy demand sector of the COAL2 model explores the interface between the energy system and the economy.

Historically, energy consumption and economic growth have been closely related. For example, GNP grew at 3.5 percent per year in the United States between 1950 and 1970, and net or final energy consumption grew at 3.2 percent per year for the same period (USDI 1972, p. 11). Certainly the relationship between energy use and economic growth may be modified by several factors. For example, a general shift toward the use of electricity, a less efficient form of energy, has caused gross United States energy consumption to grow at 3.5 percent per year from 1950 to 1970, slightly faster than net consumption. Energy price changes can also be expected to alter the relationship between energy use and GNP. Price increases will stimulate reductions in the use of energy in all sectors of the economy, and cause new shifts among the forms of energy used.

The COAL2 demand sector contains a simple structure that relates long-term changes in patterns of GNP growth and energy price behavior to the demand for energy. In this analysis, energy demand is defined as the energy used by the final consuming sectors (residential and commercial, industrial, and transportation), and therefore does not include the losses experienced in converting primary sources of energy to secondary sources (such as electricity or synthetic fuels).

The following sections of this chapter describe the historical behavior of the important variables in the demand sector, the basic

17

concepts that underlie the model formulation, and the causal structure that brings these concepts together. The last section presents a series of simulation runs which illustrate the dynamic behavior of the demand sector and examine the sensitivity of net energy demand to energy policy changes.

HISTORICAL BEHAVIOR MODES

The most prominent long-term characteristic of the United States energy system has been exponential growth in the total consumption of energy. Figure 3-1 illustrates the historical growth in gross and net energy consumption, increasing at 3.5 and 3.2 percent per year, respectively. The historical correspondence between changes in the growth of GNP and energy use has been very close, as demonstrated by Figure 3-2. Figure 3-3 shows the behavior of energy use/GNP through time. Most analysts agree that there is a slight long-term downward trend in the behavior of gross and net energy consumption per constant dollar gross national product, due to a long-term increase in the efficiency of the end uses of energy.

As total consumption of energy has grown in the United States, significant shifts in the patterns of energy use have occurred. Figure

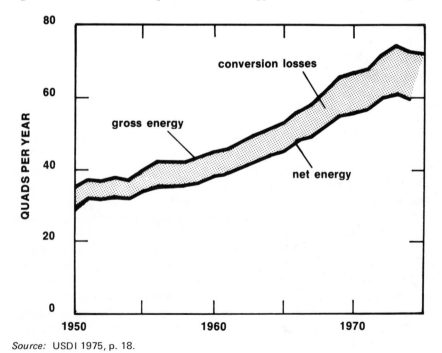

Source: USDI 1975, p. 18.

Figure 3-1.   Gross and net energy consumption

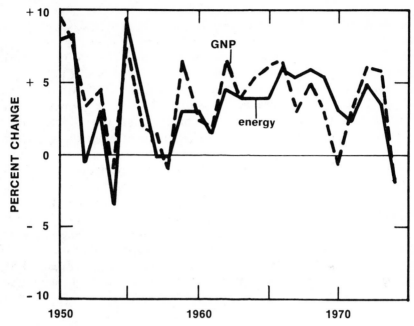

Source: USDI 1975, p. 21.

Figure 3-2. Percent change in energy use and GNP

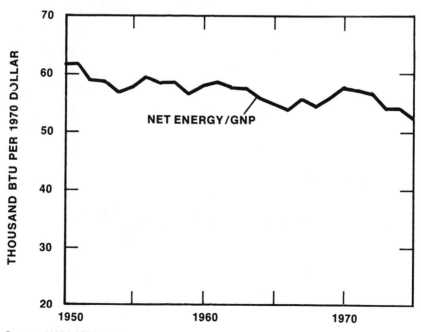

Source: USDI 1972, p. 6.

Figure 3-3. Net energy consumption per dollar of GNP

3-4 shows these shifts in the composition of United States net energy consumption over time. Oil and gas have comprised by far the largest fraction of net energy inputs, increasing to over 80 percent in 1975. The direct use of coal has declined from 36 percent of net energy consumption in 1950 to less than 8 percent in 1975. Although the use of electricity accounted for only 11 percent of net energy consumption in 1975, electricity use has grown at over 7 percent per year from 1950 to 1972. Because net energy consumption has grown at only 3.2 percent per year over the same period, electricity's share of final demand has increased rapidly, as shown in Figure 3-4.

Relative price changes are often the cause of shifts in the patterns of use of substitutable end products. Yet Figure 3-5 shows no distinct long-term relative changes in the prices of coal, oil and gas, and thus other factors such as fuel convenience must be considered to explain coal's decline in market share. The price of electricity has dropped significantly relative to coal and oil and gas (Figure 3-6), and is at least partially responsible for the rapid shift to electricity as an end-use energy form.

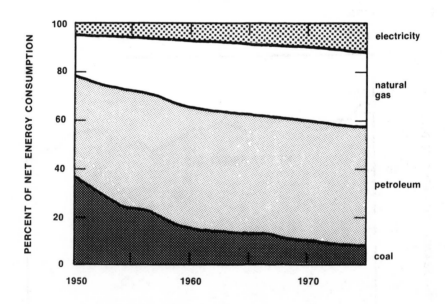

Source: Data from USDI 1975, p. 28 and USDI 1972, Appendix B.

Figure 3-4. Net energy market share history

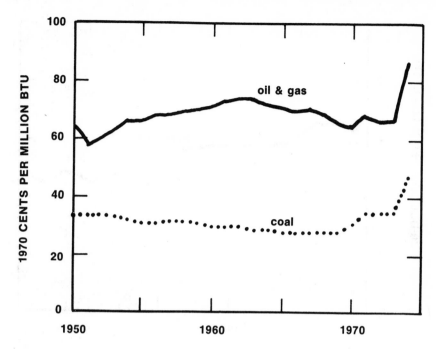

Source: Data from *Steam-Electric Plant Factors* 1975; *Gas Facts* 1974; *Minerals Yearbook* 1974.

Figure 3-5. Fuel price history

BASIC CONCEPTS

The Definition of Energy Demand

Energy demand is an elusive concept, defined differently almost as often as it is used. Energy demand forecasts are normally simple extrapolations of historical trends in energy consumption. The extrapolations are variously labeled as forecasts of energy consumption, demand, needs, or requirements, but usually little recognition is given to the conceptual differences which exist among these terms (CIIA 1972, p. 5). In the COAL2 model, energy demand represents the amount of energy desired or preferred by the population under a given situation. As energy prices and the GNP change, the situation under which desires or preferences are determined changes, and energy demand may change. *Energy demand* is defined as the amount and types of energy desired by the population under varying circumstances of energy prices and income levels (as measured

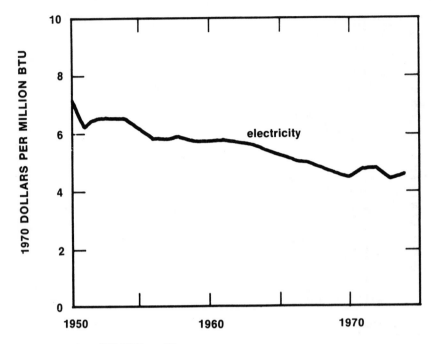

Source: Data from EEI 1974a, p. 53.

Figure 3-6.   Electricity price history

indirectly by GNP). These desires may not be completely satisfied, as during the Arab oil embargo of 1973.

Energy demand is often determined by correlating the growth in *gross* energy consumption with GNP (for example, Figure 3-2). But the gross consumption of energy represents the consumption of the energy inputs (coal, oil, gas, uranium) used to produce the final energy products (electricity and processed coal, oil, and gas) consumed by the industrial, transportation, household and commercial sectors of the economy. At the point of consumption, the purchasers of energy have little concern for how the energy was processed to reach the market. Energy consumers simply respond to their income, tastes, and the price of the final energy products when purchasing various energy forms. Therefore, the energy demand sector of COAL2 forecasts the behavior of *net* or final energy demand, after conversion, processing, and delivery. Because the conversion of energy is explicitly represented in the oil and gas, electricity, and coal sectors of COAL2, these sectors forecast the gross energy demand for primary energy sources as derived from the net demands of the energy demand sector.

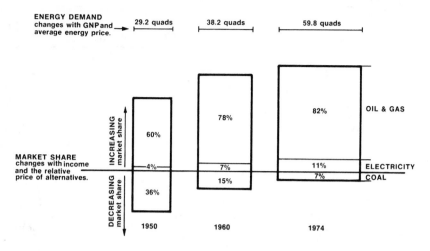

Figure 3-7. Dynamics of net energy demand

Dynamic Mechanisms of Net Energy Demand

The COAL2 model differentiates between three final energy products: oil and gas (one product), electricity, and coal. The basic purpose of the energy demand sector is to simulate long-term changes in the amount of, and preferences among, these three products in response to changes in GNP and prices. Figure 3-7 illustrates this process: changes in GNP and the average price of energy determine the long-term behavior of net energy demand. The relative inconvenience of coal as a direct fuel has decreased its share of the final energy market in the past, for shifts towards more convenient energy-consuming devices (cars, diesel locomotives, oil burners) have virtually excluded the direct use of coal. This shift away from direct coal use is included in the COAL2 model as an income effect: as GNP rises, consumers can afford the more convenient forms of fuel. Changes in the relative price of electricity and increasing incomes have accounted for the increasing fraction of net energy demanded as electricity. The remaining bulk of net energy (80 percent in 1975) is demanded as oil and gas. Once the net demands for oil and gas, electricity, and coal are determined, each producing sector allocates capital, labor, and resources to meet the consumer desires.

In many energy studies the demand for final energy products is often further disaggregated among energy-consuming sectors to obtain greater forecasting accuracy. Yet if the end users of energy are divided into three groups—(1) industrial, (2) transportation, and

Year	Consumer Group	Fuel Type			Total Group All Types
		Coal	Oil & Gas	Electricity	
1950	Industrial	20.0%	21.5%	1.9%	43.5%
	Transportation	5.7	23.3	0.1	29.1
	Residential & Commercial	9.8	15.8	1.8	27.4
	Total type, all groups	35.6	60.6	3.8	100.0
1960	Industrial	12.2	26.1	3.4	41.7
	Transportation	0.2	28.1	0.0	28.4
	Residential & Commercial	2.6	24.0	3.3	29.9
	Total type, all groups	15.0	78.2	6.8	100.0
1974	Industrial	7.0	28.7	4.5	40.2
	Transportation	0.0	30.5	0.0	30.5
	Residential & Commercial	0.5	22.6	6.2	29.3
	Total type, all groups	7.5	81.8	10.7	100.0

Source: Data from USDI 1972 and USDI 1975.

Figure 3-8.   Net energy use by consumer group and fuel type

(3) household and commercial—as in Figure 3-8, one notes that the fraction of net energy used by each group has changed very little over the period 1950-1974. The change in the groups' demands for energy shown in Figure 3-8 can be summarized as follows:

Coal has lost ground in all markets. Its share of the final energy market was 36 percent in 1950, and only 7.5 percent in 1974. The transportation and household/commercial markets have almost totally ceased using coal.

Oil and gas have increased their share of the final energy market from 60 percent in 1950 to 82 percent in 1974. The distribution of use between the three user groups has not changed significantly. Each has increased its share proportionally.

Electricity has increased its fraction of the total market from 3.8 percent in 1950 to almost 11 percent in 1974. The use of electricity is divided evenly between the industrial and the household/commercial sectors. Very little electricity is used for transportation.

Because the long-term trends away from coal and toward electricity and oil and gas are consistent across the energy-consuming sectors, no new insights would be gained by disaggregating energy demand among energy-consuming groups.

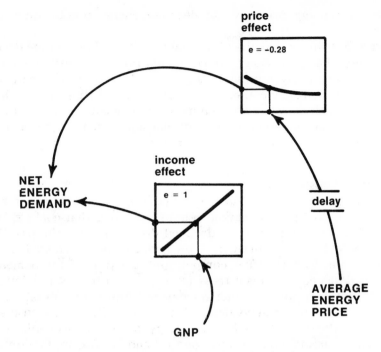

Figure 3-9. Net energy demand mechanism in COAL2

Determination of Total Energy Demand

Figure 3-9 illustrates the causal structure that determines net energy demand in the COAL2 model. Final demand for energy is determined by GNP and the average energy price. GNP is exogenously determined in the model. The average energy price is calculated by weighting the prices of oil and gas, electricity, and coal with their usage rates.

The responsiveness of energy demand to a change in price is often measured by a price *elasticity*. (For a further discussion of price elasticity, see, for example, Watson 1963.) As energy prices increase, the efficiency of energy use per dollar of GNP can be expected to increase as conservation measures are implemented throughout the economy. The elasticity of total demand measures the percentage change in final demand caused by a 1 percent change in price. A number of studies have estimated the long-run elasticity of total demand to be between —0.15 and —0.5. (M.I.T. 1974, p. 29: e = —0.15; FEA 1976: e = —0.3 to —0.5; COMRATE 1975a, p. 311: e = —0.25). Our own analysis produced a figure of —0.28, well within this range. A price elasticity of —0.28 implies that a 50 percent increase in energy prices would decrease energy use by only 14 percent.

Figure 3-9 shows that the COAL2 model also includes a time delay before any change in energy price has its full effect on the energy consumer. The delay represents the time taken to perceive and act upon a price change by conserving energy, or by refitting or replacing equipment with new energy-conserving equipment. In the COAL2 model the average response time of consumer demand to energy price changes is set at 10 years (Chapman 1972, p. 705).

Interfuel Substitution

Oil and gas, coal, and electricity are the major competitors for a share of the final energy market. Figure 3-4 shows that coal's share of the final energy market has declined drastically over the past 25 years, while both electricity and oil and gas have increased their market share. The interfuel competition structure of the demand sector is designed to explain these trends, and to project the future mix of final demand. Note that the demand sector controls only the competition for *final* energy demand: the competition between fuels for the production of these final energy forms (for example, the competition among nuclear, coal, and oil and gas for production of electricity) is represented in the energy-producing sectors (Chapters Four through Six).

Figures 3-10 and 3-11 illustrate the structural assumptions controlling interfuel substitution in the model. The market shares for electricity and coal are determined by consumer income (measured by GNP) and their price relative to the major alternative, oil and gas. The oil and gas market share is what is left over after electricity's share and coal's share have been deducted. As GNP increases, consumers tend to purchase less coal and more electricity, as shown in the left-hand graphs of Figures 3-10 and 3-11. Increases in the price of each energy product (relative to its main competitor, oil and gas) tend to decrease its market share, as shown in the right-hand graph of the two figures. As in the determination of total demand, the model structure includes a 10-year delay in the adjustment of market shares to price changes.

Within the historical range of operation, the relationships shown in Figures 3-10 and 3-11 were estimated using historical data and elasticity estimates from market share studies conducted at Oak Ridge and the Federal Energy Administration (Chapman 1972, p. 705; FEA 1976, p. C-7). Extreme values of the functions in Figures 3-10 and 3-11 were determined by market saturation limits of each of the fuels. Under extreme price incentives it was estimated that the direct use of coal could satisfy a maximum of 25 percent of

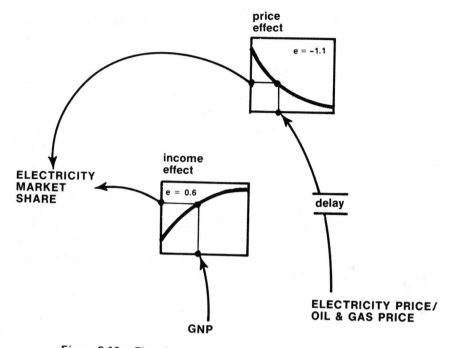

Figure 3-10. Electricity's share-of-demand mechanism in COAL2

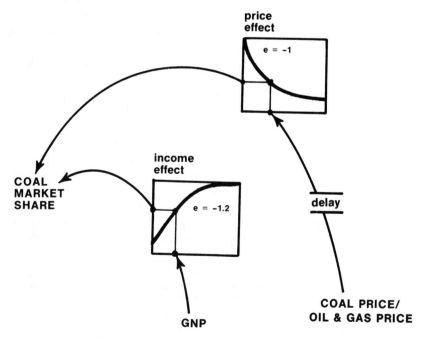

Figure 3-11. Coal's share-of-demand mechanism in COAL2

the final energy market, primarily as an industrial fuel.[a] Electricity's maximum market share was set at 70 percent, which is reached only when incomes are high and electricity prices are low (Gouse 1973, p. VI-6).

The income and price effects shown in Figures 3-10 and 3-11 can operate in the same direction, or can effectively cancel each other out by acting in opposite directions. Over the historical period both the income and the price effect increased electricity's market share. Although increases in income have caused a major shift away from the direct use of coal in the past, coal prices have decreased relative to oil and gas prices since 1973. Continued oil and gas price increases could reverse the historical trend, increasing coal's market share of net energy demand in the future. Indirect uses of coal (conversion of coal to oil, gas, or electricity) could increase United States dependence on coal well above the 25 percent limit which operates in the final energy market. The potential for coal conversion is discussed in Chapters Four and Five.

DEMAND SECTOR CAUSAL STRUCTURE

The structural assumptions described in the previous section can be combined to form a *causal diagram* of the COAL2 demand sector (Figure 3-12). In a causal diagram, system interactions are shown by arrows leading from each element to all other variables that are influenced by changes in that element.

The causal diagram in Figure 3-12 indicates that the demand sector of COAL2 has no major endogenous feedback loops. GNP, a primary determinant of both net demand and the market share for the three energy forms, is set exogenously in the model.[b] Energy prices, also important in future projections of energy demand and fuel mix components, are calculated in the energy production sectors of the model. The relationships in the demand sector form part of a set of negative feedback loops between demand and price: as energy demand increases, the average energy price tends to increase as well, thereby reducing further growth rates in demand. Energy demand increases also result in rapid increases in oil and gas prices, forcing demand to shift to other energy forms: coal and electricity. These

[a]Industrial energy use represents 40 percent of final demand (Figure 3-8). A substantial amount of industrial energy use is electrical use and feedstocks, leaving only 25 percent (direct heat and process steam in industry) available to be satisfied directly by coal (see Ross and Williams 1975, p. 19).

[b]A later version of the COAL2 model, tentatively titled ENERGY1, will close the causal link between energy and economic growth.

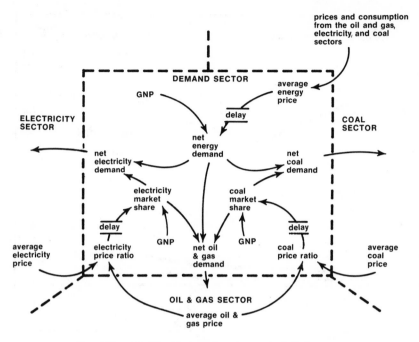

Figure 3-12. Demand sector causal diagram

feedback loops are closed in the model when the demand sector is linked with the three energy production sectors (see Chapter Seven).

DEMAND SECTOR SIMULATION RUNS

The energy demand sector of COAL2 projects future final demands for oil and gas, electricity, and coal to the year 2010. Most previous forecasts have simply extrapolated past energy trends. With the rapid changes in current and projected energy prices such forecasts are no longer adequate. This section examines the response of net energy demand and its components to changes in both GNP and government policy actions. Energy prices, normally determined in the energy producing sectors of the COAL2 model, are set exogenously in the following simulations.

Demand Sector Historical Behavior

Figure 3-13a plots the historical behavior of GNP and energy prices. From 1950 to 1973 United States GNP rose from 480 billion dollars to 1.1 trillion dollars (in 1970 dollars). Electricity prices, significantly higher than the prices of oil and gas or coal, declined until 1970. Figure 3-13a shows that the price of oil and gas and the

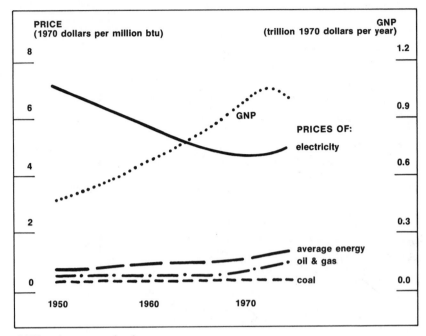

a. GNP and energy price inputs

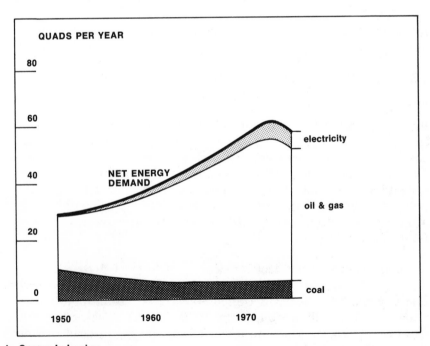

b. Sector behavior

Figure 3-13. Demand sector historical behavior

price of coal have remained constant in real terms over most of the historical period. Because of the shift to more expensive fuels (primarily electricity), the average energy price increased slightly over the historical period.

When the historical GNP and energy price data of Figure 3-13a are used as inputs to the demand sector, the sector behaves as shown in Figure 3-13b. Net energy demand grows from 30 quads per year in 1950 to 59 quads per year in 1975. Coal's share of final demand falls from 35 percent in 1950 to 10 percent in 1975 as increased consumer income allows the consumer to choose the convenience of oil, gas, and electricity over the inconvenience of coal. Electricity's market share increases from 3.5 percent to 11 percent over the same period, as consumers prefer its ease of use (modeled as an income effect) and its declining relative price. The bulk of the final energy market belongs to oil and gas, whose market share grows from 60 percent in 1950 to over 80 percent in 1970. After 1970, oil and gas's market share begins to decline slightly as electricity usage climbs.

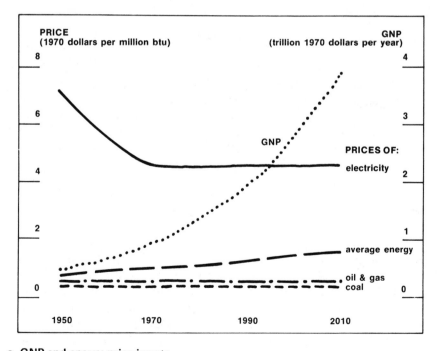

a. GNP and energy price inputs

Figure 3-14. Demand sector historical growth projection

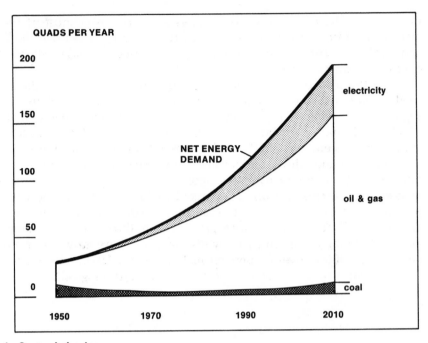

b. Sector behavior

Figure 3-14 (continued). Demand sector historical growth projection

Historical Growth Projection

Before the Arab oil embargo of 1973, a common assumption for the projection of future energy demand was that both energy prices and economic growth (GNP) would continue their historical trends. In 1972 the National Petroleum Council projected that net demand would rise to 95 quads per year by 1985 with constant future energy prices, a growth rate of 3.5 percent per year (NPC 1972, p. 37). Also in 1972, Dupree and West of the U.S. Department of the Interior made a similar projection, estimating that net demand would grow to 90 quads per year by 1985 and 140 quads per year by the year 2000 (USDI 1972, p. 2).

To compare COAL2's behavior with pre-embargo demand projections, the demand sector was run with the inputs shown in Figure 3-14a. Energy prices are assumed to remain constant in the future, and GNP continues to grow at its long-term historical rate of 3.5 percent per year (note that neither of these assumptions has held true since 1972: this is therefore a hypothetical run). The assumption of constant energy prices implies a continuation of the increase in the *average* energy price (Figure 3-14a), since a greater fraction of electricity, a more expensive form of energy, will be consumed in the future.

Figure 3-14b shows that in the historical growth projection, net energy demand grows to 92 quads per year in 1985 and 148 quads per year in 2000. With constant future energy prices, the historical trends in end-use fuel market shares tend to continue. Coal's market share drops from its 1975 share of 8 percent to 5 percent of final energy demand by 2000. Electricity demand increases from 11 percent in 1975 to 20 percent by the turn of the century. Because of the large increase in electricity use, oil and gas demand drops from 80 percent of net demand in 1970 to 75 percent in the year 2000. Since these trends are consistent with the NPC and Interior studies, the behavior of the COAL2 demand sector seems plausible under historical growth conditions.

Demand Sector Reference Projection

Figure 3-15 illustrates the *reference*, or base case, projection of the COAL2 demand sector. Because the demand for energy is so intimate-

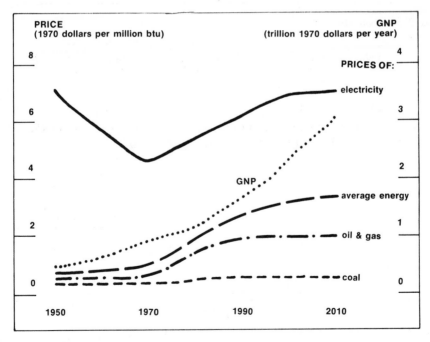

a. GNP and energy price inputs

Figure 3-15. Demand sector reference projection

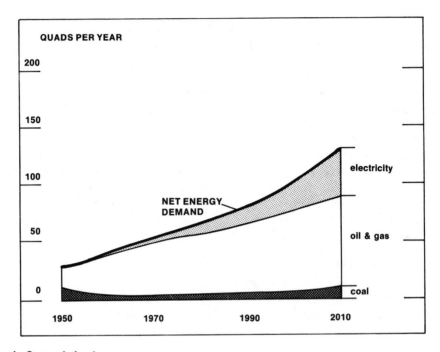

b. Sector behavior

Figure 3-15 (continued). Demand sector reference projection

ly bound to economic activity, a forecast of GNP is necessary to project energy demand. In the COAL2 reference projection, GNP was assumed to be the following (see also Figure 3-15a):

Year	GNP (trillions of 1970 dollars)	Growth Rate (average % per year from 1975)
1975	1.03*	—
1985	1.4	3.4
2000	2.3	3.2
2010	2.9	3.0

*Preliminary estimate.

Source: Data Resources, Inc., forecast as reported in FEA 1976, p. B-4.

Figure 3-16. Forecasted GNP growth

In Figure 3-16, GNP is forecast to grow at 3.4 percent per year from 1975 to 1985, slowing to a 3 percent per year average rate of growth to the year 2010. This growth rate is substantially below the long-term average of 3.5 percent per year (and the 4 percent per year average of the 1960s). The growth capacity of the United States economy is assumed to diminish during the next 35 years because of the diversion of capital to nongrowth purposes (primarily energy production and pollution control), and the falling growth rate of the adult population. Figure 3-15a also illustrates the assumed behavior of energy prices for the demand sector reference run. These prices are representative of simulations from the oil and gas, electricity, and coal sectors described in Chapters Four through Six. Oil and gas prices (in constant dollars) are projected to rise by 220 percent, coal prices by 70 percent, and electricity prices by 50 percent by the year 2010.

Figure 3-15b shows the reference projection of net energy demand and its components obtained by using the variables shown in Figure 3-15a as exogenous inputs. Net demand grows to 73 quads per year in 1985 and about 100 quads per year in 2000, an average growth rate of 2 percent per year. This projection is considerably below the historical net demand growth rate of 3.2 percent per year, due both to increases in energy efficiency (stimulated by energy price increases), and to decreases in the projected growth in GNP. As shown in Figure 3-17, the COAL2 model reference projection of energy demand is quite close to similar projections made by the Department of the Interior, FEA, and ERDA for the year 1985, and slightly lower than comparable projections in the year 2000.

Figure 3-15b also projects the final demand market shares of oil and gas, electricity, and coal to the year 2010. Coal's share of final energy demand remains near its 1975 market share (8 percent) from 1975 to the year 2010 as a result of offsetting effects from increased income (discouraging direct coal use) and higher oil and gas prices (encouraging coal use). Electricity's market share increases to 25 percent by the year 2000, slightly greater than the historical growth projection of 20 percent (Figure 3-14), due to higher oil and gas prices.

	Net Energy Demand in Quads per Year			
	COAL2 Reference Projection	*USDI 1975*	*FEA 1976 ($13/bbl.)*	*ERDA 1975*
1985	73	77	76	75
2000	102	110	–	111

Figure 3-17. COAL2 demand projection compared to other studies

		COAL2 Reference Projection	USDI 1975	FEA 1976 ($13/bbl.)	ERDA 1975
1985	Coal	9%	5%	7%	9%
	Electricity	16	17	14	16
	Oil & Gas	75	77	79	75
2000	Coal	9	5	–	14
	Electricity	25	27	–	21
	Oil & Gas	66	68	–	65

Figure 3-18. COAL2 **market share projection compared to other studies**

Figure 3-18 compares the COAL2 reference projection of energy demand market shares with the three studies listed in Figure 3-17. Here the variation among studies is more pronounced, as the different methodologies used for the projections lead to different results. Because the Department of the Interior study does not include price effects (USDI 1975, p. 25), coal's share of final demand is lower than the COAL2 reference projection. In contrast, the linear programming approach used in the ERDA study does not include the delays inherent in increasing coal's market share (ERDA 1975, p. B-1) and thus yields a higher estimate for coal in the year 2000 (14 percent of final demand, compared to 9 percent in COAL2). Trends toward increasing electricity use and away from the use of oil and gas are evident in all of the market share studies in Figure 3-18.

Accelerated Conservation Projection

Under the heading of "Accelerated Conservation," the COAL2 model examines the savings in demand both from increases in end-use efficiency (similar to the Ford "Technical Fix" and ERDA "Improved Efficiency in End-Use" scenarios) and from direct conservation measures (where an energy-consuming device is used less). While energy price increases provide a strong incentive toward efficiency and conservation, such efforts could be accelerated by government intervention. Some of the more significant improvements suggested by FEA, ERDA, and the Ford Energy Policy Project are incentives that increase the use of heat pumps, encourage better insulation, increase automobile mileage, and provide more efficient aircraft and higher aircraft load factors. The FEA estimates that net energy demand could be reduced to 70 quads per year by 1985 with accelerated conservation measures, a 7 percent reduction from their price-sensitive forecast at world oil prices of 13 dollars per barrel

(FEA 1976, pp. G-5, G-7). Both ERDA and the Ford Energy Policy Project estimate similar savings for 1985 and project major potential reductions in energy demand (approximately 20 percent) by the year 2000 if conservation programs are accelerated (ERDA 1975, pp. B-12, B-13; Ford 1974a, pp. 21, 46, 99).

As discussed in the original FEA Project Independence report (FEA 1974, p. 158), the primary justification for government intervention to accelerate energy conservation would be to achieve the national goal of independence from foreign oil supplies, a goal not reflected in the normal marketplace decisions controlling energy supply and demand. To model the effects of such conservation measures, it is assumed that government incentive programs enhance the normal market response to increasing energy prices included in the COAL2 demand sector. (For a description of possible government policy options, see Ford 1974a, p. 45; FEA 1976, p. E-6.) Figure 3-19 illustrates the increase in the price elasticity of demand attributable to accelerated conservation. Assuming that the average energy price increases by a factor of 2 by 1985 over the 1970 value and a factor of 3 by the year 2000 (see Figure 3-15), the FEA, ERDA, and Ford Energy Policy Project demand reductions imply an

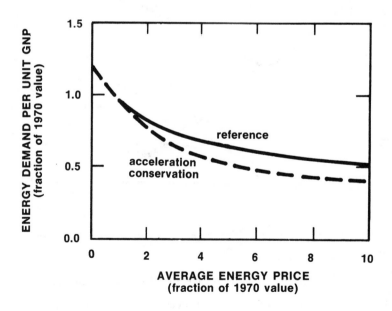

Figure 3-19. Effects of Accelerated Conservation policies on demand price elasticity

average increase in the price elasticity of total demand from its reference value of −0.28 to a new value of −0.40 (see Figure 3-18).[d]

Figure 3-20 shows the behavior of the COAL2 demand sector with Accelerated Conservation. The price and GNP inputs (Figure 3-20a) are identical to the reference run, Figure 3-15. Yet because additional conservation measures are instituted as the average energy price rises, net energy demand increases to only 70 quads per year by 1985 and 90 quads per year by the year 2000, an average growth rate of 1.7 percent per year. Although the demand reductions attributable to the Accelerated Conservation policies are considerable, energy demand continues to grow through the energy transition period.

Zero Energy Growth Projection

The Ford Energy Policy Project has suggested a second policy strategy that attempts to stabilize energy demand over the long

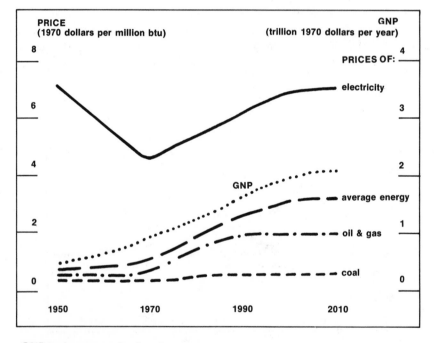

a. GNP and energy price inputs

Figure 3-20. Demand sector Accelerated Conservation projection

---

[d]For example, ERDA-48 projects a decrease in energy demand by 2000 to 90 quads per year due to accelerated conservation (ERDA 1975, pp. B11, B13). This is 65 percent of 140 quads per year, the demand that would be reached by 2000 if prices remained constant (see Ford 1974a, p. 21; USDI 1972; Figure 3-14 in Chapter Three of this book). Assuming a factor of 3 price increase, the price elasticity of demand is: $1n(.65)/1n(3) = −0.40$.

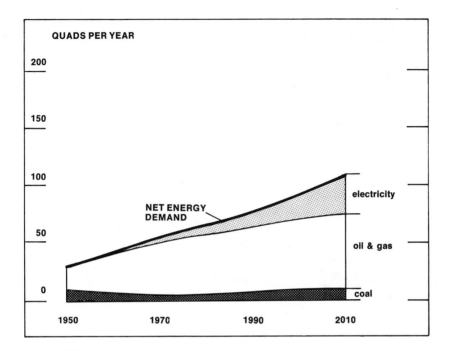

QUADS PER YEAR

NET ENERGY
DEMAND

electricity

oil & gas

coal

1950   1970   1990   2010

b. Sector behavior

Figure 3-20 (continued). Demand sector Accelerated Conservation projection

term—Zero Energy Growth (ZEG). United States Bureau of the
Census projections indicate that zero population growth could be
reached as early as 2020 if current fertility trends continue (Bureau
of the Census 1972). A transition to a stable energy-consuming
"post-industrial" society with an emphasis on services rather than
material goods production may take equally as long. In the demand
sector of COAL2, energy demand increases proportionally with
material growth (GNP), given constant energy prices. An optimistic
timetable for an orderly transition to an equilibrium society might be
a goal that stabilizes the growth inducements from GNP by the year
2010.

The policy instruments that achieve a shift to Zero Energy Growth
are discussed in a number of sources (Meadows 1972; Blueprint
1972; Daly 1973) and can be tested in COAL2 by reducing the
growth in GNP. In order to test the effects of a ZEG policy on the
energy demand sector, GNP growth was reduced gradually to zero by
the year 2010 (Figure 3-21a), in addition to the Accelerated
Conservation policy change shown in Figure 3-19.

Figure 3-21b shows the effects of a long-term strategy of ZEG on
the behavior of the energy demand sector. Net energy demand
actually levels off in the year 2000 (10 years before GNP) due to the

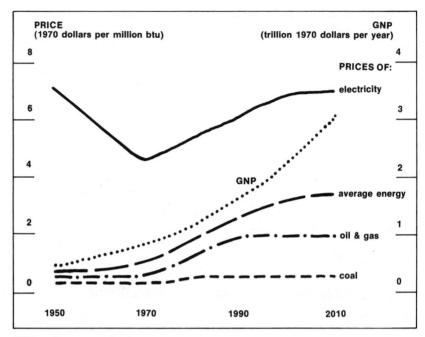

a. GNP and energy price inputs

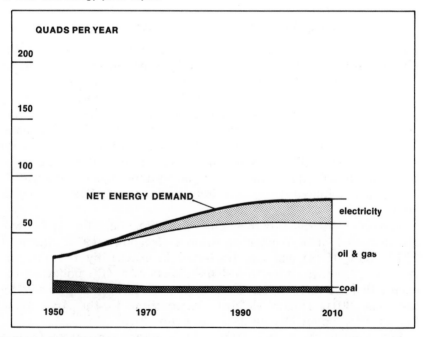

b. Sector behavior

Figure 3-21.   Demand sector Zero Energy Growth Projection

increasing use of energy-efficient technologies and conservation measures from the Accelerated Conservation policies. Energy demand is virtually unchanged from the Accelerated Conservation run through 1985 because the ZEG policies have very little effect on GNP in the short term. However, by 2010 net energy demand grows to only 80 quads per year with the ZEG policies, a 30 percent reduction from the Accelerated Conservation projection of Figure 3-20. Furthermore, the ZEG scenario remains stable after the year 2010. In contrast, if technological policies alone are implemented, energy demand will continue to grow indefinitely, continually adding to the problems of supply.

CONCLUSIONS

Figure 3-22 illustrates the four COAL2 demand projections. The *historical growth* projection assumes a continuation of past patterns of development of GNP and constant energy prices. Energy demand continues to grow at a steady 3.2 percent per year, equal to its historical growth rate. Because of the actual and projected increases in energy prices and the concurrent reduction in the rate of growth in GNP, the historical growth projection is purely hypothetical. Under any circumstances, future energy demand should be significantly reduced below historical growth rates.

If the energy demand sector is allowed to develop in a "business as usual" sense, with no major government initiatives to accelerate the

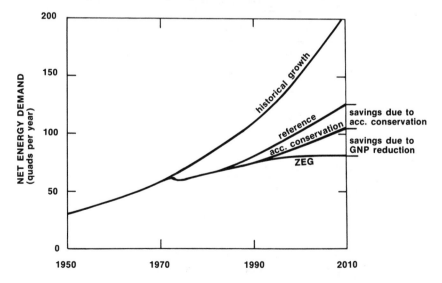

Figure 3-22. COAL2 demand projections

normal market mechanisms that reduce energy demand, the *reference projection* shown in Figure 3-22 should result. Increased prices and reduced GNP growth substantially lower the rate of growth in demand to an average of 2 percent per year over the transition period (1970-2010).

An active government program to put energy-saving technologies into use and accelerate conservation could reduce net energy demand even further below the reference projection. The *Accelerated Conservation* projection decreases the average growth rate of net energy demand to 1.7 percent per year from 1975 to 2000 (Figure 3-22). Yet these policies have only a transient effect on demand. As energy prices begin to stabilize after the year 2000, demand begins to grow at an increasing rate due to the continued growth in GNP.

Energy demand is stabilized by the end of the century in the *Zero Energy Growth* projection. In this projection, technological changes adopted in the Accelerated Conservation projection are supplemented with a long-term strategy designed to stabilize material consumption in the United States. Net energy demand levels off at 80 quads per year in Figure 3-22, 40 percent above the 1975 energy demand of 59 quads per year.

Figure 3-23 compares the COAL2 demand projections with projections made by FEA, ERDA, the U.S. Department of the Interior, and the Ford Energy Policy Project. The COAL2 reference projection is

		Projected Net Demand in Quads per Year				
Scenario	*Date*	*COAL2*	*USDI 1975*	*FEA ($13/bbl.)*	*ERDA 1975*	*FORD 1974a*
Reference Projection	1985	73	77	76	75[a]	—
	2000	102	110	—	111[a]	—
Accelerated Conservation	1985	70	—	70	67[b]	76[c]
	2000	91	—	—	89[b]	104[c]
Zero Energy Growth	1985	70	—	—	—	73[d]
	2000	80	—	—	—	80[d]

[a]ERDA scenario 0.

[b]ERDA Scenario I.

[c]Ford Technical Fix.

[d]Ford ZEG.

Figure 3-23. COAL2 projections compared to other studies

slightly below the Department of the Interior and ERDA projections in 2000, primarily due to its assumed reduction in GNP growth to an average of 3.2 percent per year. The COAL2 Accelerated Conservation projection is quite similar to the ERDA-48 "Improved Efficiencies in End-Use" scenario, while slightly lower than the Ford "Technical Fix" scenario, a result of the latter's exclusion of the effects of the 1973-1974 recession. The COAL2 ZEG projection is very similar to the Ford Foundation ZEG projection.

While the COAL2 demand sector projects the *net* or final demand for oil and gas, electricity, and coal, most energy studies report the results of their projections in terms of *gross* demand or consumption. COAL2 projects gross energy demands when the demand sector is connected to the three energy supply sectors which account for energy conversion losses. Since large conversion losses occur in the generation of electricity and synthetic fuels, gross energy consumption could rise even if net demand remains constant as electricity becomes still more popular and synthetic fuels become a factor. The interactions between energy consumers and producers are examined in Chapter Seven when all sectors of the COAL2 model are run together.

CHAPTER 3
EQUATIONS

```
100 *       ENERGY DEMAND SECTOR OF COAL2
110 NOTE
120 NOTE    TOTAL ENERGY DEMAND
130 NOTE
140 A       NED.K=EGNPR70*GNP.K*DMP.K
150 C       EGNPR70=5.77E4
160 L       GNP.K=GNP.J+(DT)(GNPIR.JK)
170 N       GNP=GNPI
180 C       GNPI=4.81E11
190 R       GNPIR.KL=GNP.K*GNPGR.K
200 A       GNPGR.K=CLIP(GNPGR1.K,LTGR.K,TIME.K,RSYEAR)
210 C       RSYEAR=1973
220 A       GNPGR1.K=CLIP(LTGR.K,RYGR.K,TIME.K,RCYEAR)
230 C       RCYEAR=1975
240 A       LTGR.K=TABLE(LTGRT,TIME.K,1950,2010,10)*1E-2
250 T       LTGRT=3.55/3.55/3.55/3.4/3.2/3/2.8
260 A       RYGR.K=TABHL(RYGRT,TIME.K,1974,1976,1)*1E-2
270 T       RYGRT=-2.1/-3.6/3.5
280 A       DMP.K=SMOOTH(IDMP.K,DAT)
290 N       DMP=1.09
300 C       DAT=10
310 A       IDMP.K=CLIP(IDMP2.K,IDMP1.K,TIME.K,PYEAR)
320 A       IDMP1.K=TABHL(IDMP1T,AEP.K/AEPN,0,10,1)
330 T       IDMP1T=1.2/1/.82/.74/.68/.64/.61/.58/.56/.54/.52
340 A       IDMP2.K=TABHL(IDMP2T,AEP.K/AEPN,0,10,1)
350 T       IDMP2T=1.2/1/.82/.74/.68/.64/.61/.58/.56/.54/.52
360 C       AEPN=.94E-6
370 A       AEP.K=(AOGP.K*NOGD.K*OGCDR.K+EP.K*TEG.K+
380 X       CPRICE.K*DCUD.K*CPDR.K)/NEC.K
390 A       NEC.K=NOGD.K*OGCDR.K+TEG.K+DCUD.K*CPDR.K
400 NOTE
410 NOTE    INTERFUEL SUBSTITUTION
420 NOTE
430 A       FEDC.K=TABHL(FEDCT,GNP.K/GNP70,.5,1.9,.2)*CDSM.K
440 T       FEDCT=.35/.15/.105/.087/.07/.06/.055/.05
450 C       GNP70=974E9
460 A       CDSM.K=TABLE(CDSMT,SCOPR.K/SCOPR70,0,2,.2)
470 T       CDSMT=5/4.5/2.5/1.7/1.3/1/.83/.71/.63/.56/.5
480 C       SCOPR70=.52
490 A       SCOPR.K=SMOOTH(COPR.K,DAT)
500 N       SCOPR=.54
510 A       COPR.K=CPRICE.K/AOGP.K
520 A       DCUD.K=FEDC.K*NED.K
530 A       FEDE.K=TABHL(FEDET,GNP.K/GNP70,0,8,1)*EDSM.K
540 T       FEDET=.03/.093/.14/.18/.21/.24/.26/.275/.28
550 A       EDSM.K=TABLE(EDSMT,SEPR.K/SEPR70,0,2.5,.25)
560 T       EDSMT=2.5/1.9/1.5/1.22/1/.78/.64/.56/.5/.44/.4
570 C       SEPR70=8.75
580 A       SEPR.K=SMOOTH(EPR.K,DAT)
590 N       SEPR=13.5
600 A       EPR.K=EP.K/AOGP.K
610 A       NELD.K=FEDE.K*NED.K
620 A       NOGD.K=(1-FEDE.K-FEDC.K)*NED.K
630 NOTE
640 NOTE    EXOGENOUS INPUTS
```

```
650 NOTE
660 A       AOGP.K=TABLE(AOGPT,TIME.K,1950,2010,10)*1E-6
670 T       AOGPT=.65/.65/.65/1.5/2/2.1/2.1
680 A       CPRICE.K=TABLE(CPRICET,TIME.K,1950,2010,10)*1E-6
690 T       CPRICET=.34/.33/.35/.5/.6/.6/.6
700 A       EP.K=TABLE(EPT,TIME.K,1950,2010,10)*1E-6
710 T       EPT=7.16/5.76/4.66/5.4/6.2/7/7
720 NOTE
730 NOTE    SUPPLEMENTARY EQUATIONS
740 NOTE
750 A       ECDR.K=NEC.K/NED.K
760 A       TEI.K=DOGINV.K+SEUI.K+CCINV.K
770 A       EIGNPR.K=TEI.K/GNP.K
780 A       COGD.K=DCUD.K+NOGD.K
790 A       OGCDR.K=1
800 A       CPDR.K=1
810 A       TEG.K=NELD.K
820 A       DOGINV.K=0
830 A       SEUI.K=0
840 A       CCINV.K=0
850 NOTE
860 NOTE    CONTROL CARDS
870 NOTE
880 N       TIME=1950
890 C       PYEAR=1977
900 SPEC    DT=.5/LENGTH=0/PLTPER=2/PRTPER=5
910 PLOT    NED=D,DCUD=C,COGD=O(0,2E17)
920 PLOT    GNP=P(0,4E12)/AEP=$,CPRICE=+,AOGP=#,EP=*(0,8E-6)
930 PRINT   NED,NOGD,NELD,DCUD,FEDE,FEDC
940 RUN
950 NOTE    PARAMETER CHANGES FOR SIMULATION RUNS
960 NOTE
970 C       PLTPER=1
980 C       LENGTH=1975
990 PLOT    NED=D,DCUD=C,COGD=O(0,80E15)
1000 PLOT   GNP=P(0,1.2E12)/AEP=$,AOGP=#,CPRICE=+,EP=*(0,8E-6)
1010 RUN    HISTORICAL RUN
1020 CP     LENGTH=2010
1030 PLOT   NED=D,DCUD=C,COGD=O(0,2E17)
1040 PLOT   GNP=P(0,4E12)/AEP=$,AOGP=#,CPRICE=+,EP=*(0,8E-6)
1050 C      RSYEAR=4000
1060 T      LTGRT=3.55/3.55/3.55/3.5/3.5/3.5/3.5
1070 T      AOGPT=.65/.65/.65/.65/.65/.65/.65
1080 T      CPRICET=.35/.35/.35/.35/.35/.35/.35
1090 T      EPT=7.16/5.76/4.66/4.66/4.66/4.66/4.66
1100 RUN    HISTORICAL GROWTH PROJECTION
1110 RUN    REFERENCE PROJECTION
1120 T      IDMP2T=1.2/1/.76/.64/.57/.53/.49/.46/.44/.42/.40
1130 RUN    ACCELERATED CONSERVATION PROGRAM
1140 T      IDMP2T=1.2/1/.76/.64/.57/.53/.49/.46/.44/.42/.40
1150 T      LTGRT=3.55/3.55/3.55/3.4/2.6/1.2/0
1160 RUN    ZERO ENERGY GROWTH PROJECTION
```

✳ *Chapter 4*

The Oil and Gas Sector

PURPOSE OF THE OIL AND GAS SECTOR

The oil and gas sector of COAL2 simulates the long-term dynamics of oil and gas supply in the United States. Oil and gas can be produced domestically from two major sources: conventional oil and gas wells or synthetic production from coal. (Due to its resource and environmental limitations, shale oil is not included explicitly in COAL2.)[a] By current estimates, the United States has depleted almost 45 percent of its conventional oil and gas resources and the remaining resources can be extracted only at much higher capital expense (USGS 1975, p. 3).

As conventional resources of oil and gas are depleted, the economic incentives to shift investment to the production of synthetic sources of oil and gas will increase. However, synthetics are also expensive (cost estimates range from 2 to 2.60 dollars per million Btu in 1970 dollars, compared to pre-embargo oil prices of 65 cents per million Btu) and require a significant amount of additional research and development before they can become a proven energy source. The supply problem for oil and gas could therefore be critical during the next 25 years: conventional production could decline significantly due to resource depletion and price regulation, and synthetic fuels could be delayed until past 1990, forcing the United States to import increasing amounts of foreign oil.

---

[a]See Rattein and Eaton 1976, p. 153. Conversion of coal to synthetic fuels and the retorting of oil shale share similar economic and environmental problems. The COAL2 model makes the simplifying assumption that all synthetic fuels will be based on coal.

A number of policies have been suggested to reduce future United States dependence on foreign oil. Oil import tariffs and quotas, price deregulation, secondary and tertiary recovery of existing oil and gas reserves, accelerated development of synthetic fuels, and reduction of oil and gas demand are those policies most often suggested to alleviate the necessity for increased imports.

This chapter describes the structure, assumptions, and behavior of the COAL2 model's oil and gas sector. In the final section, the model is used to evaluate the effectiveness of the above oil and gas policy alternatives.

HISTORICAL BEHAVIOR

Petroleum and natural gas consumption in the United States has grown to supply almost 80 percent of total United States energy in 1970 (Figures 1-2 and 4-1). From the turn of the century to the mid-1960s, domestic production of oil and gas grew at approximately 6 percent per year as the United States enjoyed a period of success in exploration, lower production costs, and excess production capacity.

Yet in the ten years since the mid-1960s (one doubling time of production), the United States' abundance of oil and gas has dramatically disappeared. The domestic production of crude oil peaked in 1970 (Figure 4-2), with the 1975 preliminary production estimate indicating a 15 percent drop from 1970 levels (FEA 1976, p. 3). At the same time, Figure 4-2 shows that petroleum consumption has continued to grow (except during the recession years), with the result that petroleum imports have continually increased to over 6 million barrels per day (mbpd) in 1975, equal to 13 quads per year. Domestic production of natural gas remained essentially stable in the

Year	Total Energy Consumption (Quads/year)	Percent Oil & Gas (%)	Oil & Gas Consumption (Quads/year)	Oil (Quads/year)	Gas (Quads/year)
1950	34.0	59	19.9	13.8	6.1
1955	39.7	69	27.2	18.0	9.2
1960	44.6	75	33.5	20.9	12.6
1965	53.3	76	40.5	24.4	16.1
1970	67.1	79	53.1	31.1	22.0
1975	71.1	74	52.8	32.7	20.1

Source: USDI 1972, Appendix B. 1975 data from ERDA 1976, p. 91.

Figure 4-1. Oil and gas consumption

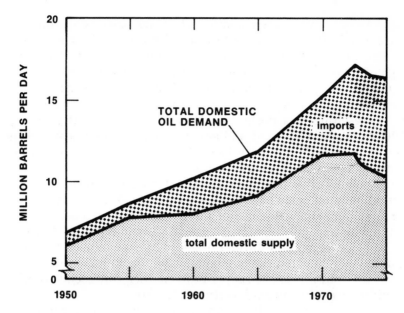

Source: Data from FEA 1976, p. xxiii.

Figure 4-2. Petroleum production and consumption

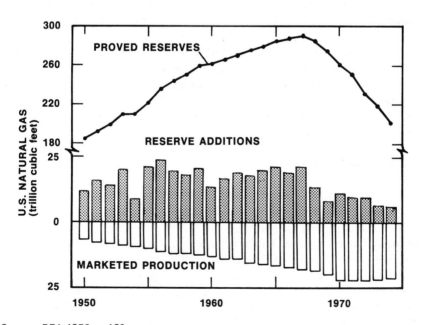

Source: FEA 1976, p. 120.

Figure 4-3. Natural gas reserves, additions, and production

early 1970s, while reserve additions have remained below production since 1967 (Figure 4-3). Preliminary 1975 data show a 10 percent decline in United States natural gas production from 1973's record high of 22.5 trillion cubic feet per year (FEA 1976, p. 8).

One direct cause of the current oil and gas shortage has been a slowdown in the growth of domestic capital investment (Figure 4-4a), resulting in a decline in domestic drilling activity during the 1960s (Figure 4-4b). Oil and gas capital investment ceased to grow during the 1960s due to a lack of economic incentives. Figure 4-4 shows a comparison of the cost and price of new oil from 1956 to 1975. While oil prices have remained constant over most of this period, new oil costs have risen dramatically from the late 1960s to 1975 (Figure 4-5). Because of the rise in domestic costs, it became cheaper to import oil than to expand domestic capacity. At the same time, price regulation further reduced the incentives to invest in domestic oil and natural gas exploration. From a financial point of view, the solution to the supply problem therefore seems straightforward: increase oil and gas investment incentives.

Yet, at the core of the problem is a second, longer-term trend: Figure 4-6 shows that the marginal returns to drilling for crude oil declined drastically during the 1950-1960 period, thereby increasing costs and contributing to the lack of economic incentives to invest. Estimating the total amount of recoverable resources from this curve, M.K. Hubbert of the United States Geological Survey argues that oil and gas production will follow "life cycle" patterns (shown in Figure 4-7), which could explain the recent peak and decline in production of both petroleum and natural gas (Figure 4-2 and 4-3). The oil and gas sector of COAL2 includes both the more immediate effects of investment incentives and the longer-term effects of resource depletion in examining the oil and gas supply problem.

BASIC CONCEPTS

Finite Oil and Gas Resources

In any long-term analysis of United States oil and gas supply, the most crucial factor is resource depletion, for the total oil and gas resource base in the United States is limited. The definition of reserves and resources used in this text follows the guidelines established by the U.S. Geological Survey (McKelvey 1973): the term *reserves* refers to economically recoverable material in identified deposits, and the term *resources* includes deposits not yet discovered as well as identified deposits that cannot be recovered given current technology or economics (deep water deposits, small

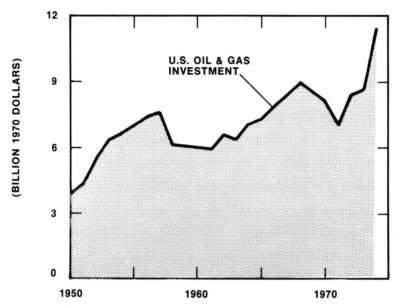

Source: Chase 1972, p. 11. Reprinted by permission.

a. Total capital investments.

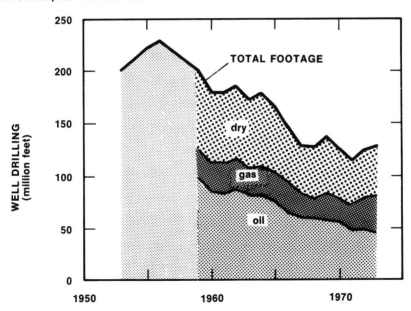

Source: Teller 1975, p. 52.

b. Oil well activity.

Figure 4-4. Oil and gas capital investment

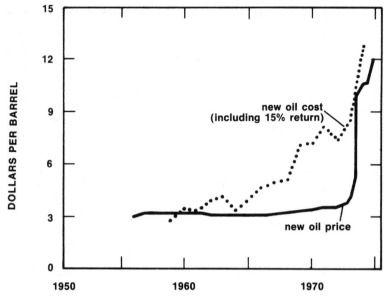

Source: Nathan 1975.

Figure 4-5. New oil cost and price

pools). Clearly, only the broader classification of oil and gas resources is in finite supply. Technological advance, price increases, and exploration activities may expand reserves for a time, but do not affect resources.

Although the total oil and gas resource base may be quite large, only a fraction (currently about 30 percent of the discovered oil and 80 percent of the discovered gas in place) is recoverable by current technologies. The analysis of the oil and gas sector of COAL2 focuses on estimates of *recoverable resources*, those resources (both discovered and undiscovered) ultimately recoverable given expected long-term developments in recovery techniques. New enhanced recovery techniques that expand the recoverable resource base beyond current estimates will be introduced in the analysis as a policy change.

While there has been little argument over the concept of finite oil and gas resources, there has been a great deal of controversy over the amount of recoverable oil and gas resources estimated to exist in the United States.[b] Within the last decades, estimates published of the total recoverable oil resource base of the United States and its adjacent continental shelves (exclusive of Alaska) have ranged from 145 to 590 billion barrels (Hubbert 1969, p. 184). Natural gas

[b]Excellent summaries of the nature and methods of the conflicting oil and gas resource estimates are included in Hubbert 1969 and COMRATE 1975a.

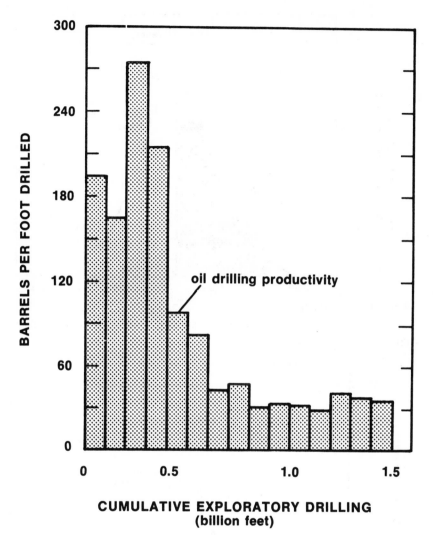

Source: Hubbert 1967, pp. 22, 23.

Figure 4-6. Oil discoveries per foot drilled vs. cumulative feet drilled

resource estimates have also varied over a wide range, since most estimates for gas are inferred from oil estimates. The estimates divide quite naturally into gwo groups: those associated with oil companies (including Hubbert), and those of the United States Geological Survey.

Past estimates by the U.S. Geological Survey tend to fall in the higher range: Zapp (1961) estimated 590 billion barrels of crude oil for the United States exclusive of Alaska; Hendricks (1965) esti-

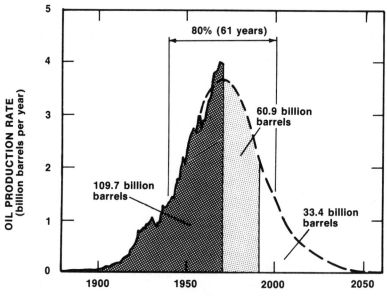

Source: Hubbert 1972, p. 198.

a. Petroleum liquids in United States and adjacent continental shelves, excluding Alaska

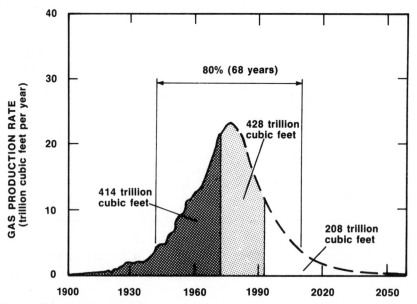

Source: Hubbert 1972, p. 152.

b. Natural gas in United States and adjacent continental shelves, excluding Alaska.

Figure 4-7. Oil and gas production life cycles

mated 338 billion barrels; and more recently, Theobald, Schwein-furth, and Duncan (1972) estimated 400 billion barrels.

The Geological Survey estimates are all based on some form of the Zapp hypothesis (Zapp 1962, pp. H-22, H-23), which assumes that the discovery characteristics of the richest deposits already selected for drilling by the oil companies are typical of entire geological basins. Thus, according to the Zapp hypothesis, the amount of oil to be produced is proportional only to drilling activity, with no importance attached to the geological variations within different parts of the basins, or to the accumulated expertise of the profes-sional, risk-taking explorers for oil and gas.

However, M.K. Hubbert has shown that the oil discovered per well foot drilled has in fact decreased drastically as a function of the cumulative drilling activity (Figure 4-8). This is in direct conflict with the Zapp hypothesis, which claims that drilling returns should remain constant (at 118 barrels/foot in Figure 4-8) over a wide range of cumulative footage drilled (to 5 billion feet in Figure 4-8). Based on such statistical projections of past drilling experience, Hubbert (COMRATE 1975a, p. 98) estimates the total recoverable oil and gas

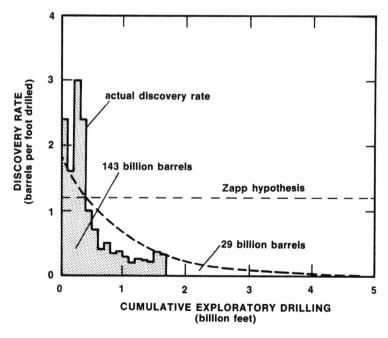

Source: Hubbert 1972, p. 125.

Figure 4-8.   Zapp hypothesis compared with actual discovery data

resources of the United States (including Alaska) to be 2,660 quads (2.66 x 10^{18} Btu). This figure corresponds closely with a recent Mobil Oil Company estimate (Mobil 1974, p. 128), and a National Academy of Sciences study (COMRATE 1975a, p. 89). Using similar resource estimation techniques, the U.S. Geological Survey has recently revised its estimates downward to correspond closely with the Hubbert, Mobil, and NAS studies (See Figure 4-9). The COAL2 model uses the Geological Survey "mean value" estimate of 270 billion barrels (1,560 quads) of petroleum liquids (oil and natural gas liquids) and 1,200 trillion cubic feet (1,240 quads) of natural gas as the total recoverable petroleum and natural gas resource base of the United States, including Alaska. By 1976, almost 50 percent of the initial recoverable oil and gas resource base had been depleted.

The oil and gas resource estimate is incorporated into the resource depletion structure illustrated in Figure 4-10. The stock of recoverable oil and gas resources is initially set to correspond with the U.S. Geological Survey estimate of 2,800 quads. As this resource stock is depleted (by the oil and gas production rate in Figure 4-10), the physical productivity of the oil and gas capital stock, measured in Btu's per year of output obtained per dollar of capital investment, is assumed to decrease. A decrease in capital productivity (equivalent

	Unit	Hubbert 1975	Mobil 1974	COMRATE 1975a	USGS 1975
*Oil**					
Recoverable beyond	10^9 bbl.	72	88	113	105
1974 known reserves	quads	420	510	660	610
Total	10^9 bbl.	235	251	276	268
Resources	quads	1,360	1,460	1,600	1,560
Gas					
Recoverable beyond	10^9 bbl.	540	443	530	489
1974 known reserves	quads	560	460	550	500
Total	10^9 bbl.	1,258	1,161	1,248	1,207
Resources	quads	1,300	1,200	1,290	1,240
Total					
Oil and gas resources	quads	2,660	2,660	2,890	2,800

*Oil includes LNG.

Sources: Hubbert 1975: Referred to in COMRATE 1975a, p. 98; Mobil 1974, p. 128; COMRATE 1975a, p. 98; USGS 1975, p. 4. Indicated and inferred reserves not included—their additions tested as a policy of extensive secondary and tertiary recovery in Chapter Five.

Figure 4-9.   Comparison of oil and gas resource estimates.

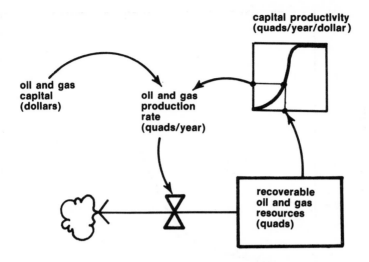

Figure 4-10. Oil and gas resource depletion structure

to a rise in capital costs) occurs as oil and gas production shifts to smaller, less productive pools or less accessible drilling locations such as offshore deposits and Alaska. Because the real capital costs of exploratory drilling have remained relatively constant over history (Joint Association Survey 1974), the data on historical returns to drilling shown in Figure 4-6 can be used as a measure of capital productivity. Figure 4-11 replots the historical data shown in Figure 4-6 as a function of the fraction of oil and gas resources remaining, and indicates the implied oil and gas capital productivity relationship used in the COAL2 model.

The oil and gas depletion structure shown in Figure 4-10 forms a negative feedback loop that tends to drive the oil and gas supply industry out of balance. Resource depletion constantly increases the marginal cost of production, making it increasingly difficult for producers to meet demand, even if demand remains constant. Producers attempt to offset the effects of resource depletion by investing greater amounts of capital in drill rigs and associated equipment, given the proper investment incentives. As conventional oil and gas costs rise, investment will eventually shift to alternative methods of oil and gas production—primarily synthetic fuels from coal.

Synthetic Fuel Technologies

The future commercialization of synthetic fuel technologies depends on both technological and economic factors in the COAL2 model. Figure 4-12 illustrates the part of the oil and gas sector

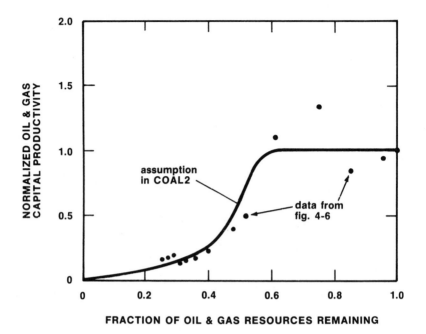

Source: Data from Figure 4 and Hubbert 1969.

Figure 4-11.   Derivation of oil and gas capital productivity relationship

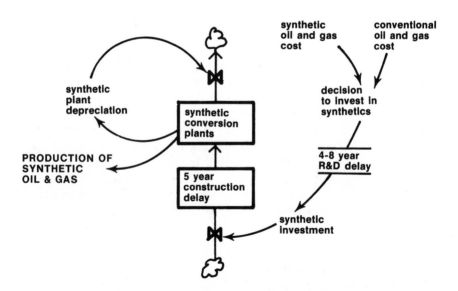

Figure 4-12.   Synthetic fuel development structure

structure that controls the development of synthetic fuels. The decision to invest in either conventional or synthetic sources of oil and gas is based on a comparison of their relative production costs. Yet even at 1976 oil prices (1.60 dollars per million Btu for imported oil in 1970 dollars), synthetic conversion facilities are not an attractive investment. Synthetic fuel costs have been estimated at approximately 2.30 dollars per million Btu (Figure 4-13), while domestic oil costs have risen to only 0.80-1.00 dollar per million Btu by 1976 (See Figure 4-5).

Even if a comparison of production costs indicated that synthetic investments are justified, construction of commercial synthetic facilities may not be technologically feasible. Synthetic fuel processes are still in their pilot-stages, requiring from 4 to 8 years of additional research and development before commercial plants are proven reliable (FEA-STF 1974, p. 40). Construction of each plant may take 5 to 6 additional years after the first demonstration plant is built (M.I.T. 1974, p. 52). Further delays (not shown in Figure 4-12) might be introduced in the acquisition of capital for new plant investments.

Figure 4-12 indicates that unless these delays are circumvented, the construction of the *first* commercial synthetic plant will likely not occur until 1990 or later. COAL2 includes the ability to examine policy options that accelerate the development of synthetic fuels by reducing R&D delays and encouraging investment in commercial plants (price guarantees). These policies are tested in the last section of this chapter.

	Cost (1970 dollars/10^6 Btu)
Feedstock	1.00
Conversion	1.14
Refining, transportation, marketing	0.14
Total delivered cost	2.28

Feedstock cost: Coal delivered at 60 cents per 10^6 Btu (approximately $14 per ton in 1970 dollars). *Conversion cost:* 270 million cubic-feet-per-day Lurgi plant. Conversion efficiency assumed to be 60% (MIT 1974, p. 57). Fixed and variable capital costs annualized at the historical average rate of 17.5 percent per year. *Refining, etc.:* Historical average (Chase 1974, p. 18).

Figure 4-13. Cost of synthetic oil or gas

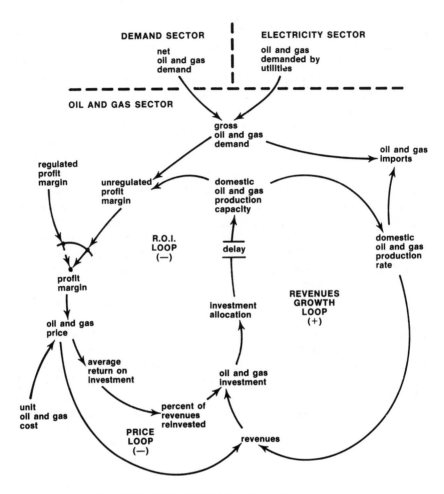

Figure 4-14. Oil and gas financial structure

Oil and Gas Financing

Figure 4-14 diagrams the feedback loop structure that controls the generation of new oil and gas investment in the COAL2 model. The ability of the oil and gas industry to finance new oil and gas investments is dependent on two factors: (1) a revenue flow that provides the main *source* of capital funds, and (2) the average return on investment, which provides the *incentive* to commit available funds to new projects.

The COAL2 model includes both internal and external financing sources. Internal funds (retained earnings, depreciation, and amortization) are calculated from revenues. Historically, internal financing has remained relatively stable, at about 30 percent of total revenues (Figure 4-15). It is likely that an equal amount could be externally

Year	Total	Investment (billion $/year) External	Internal	Revenues (billion $/year)	External/ Internal	Internal/ Revenues	Total Inv./ Revenues
1962	5.4	0.4	5.0	15.4	0.08	0.32	0.35
1963	5.2	0.5	4.7	15.8	0.11	0.30	0.33
1964	5.8	1.4	4.4	15.4	0.32	0.29	0.38
1965	5.9	1.0	4.9	15.4	0.20	0.32	0.38
1966	6.4	1.9	4.5	16.2	0.42	0.28	0.40
1967	6.8	2.8	4.0	17.5	0.70	0.23	0.39
1968	7.7	3.6	4.1	17.9	0.88	0.23	0.43
1969	7.6	2.6	5.0	17.3	0.52	0.29	0.44
1970	7.6	1.7	5.9	18.0	0.29	0.33	0.42
1971	6.8	2.6	4.2	19.1	0.62	0.22	0.36

Source: Investment: Hass, Mitchell and Stone 1974, pp. 38, 39. *Revenues:* Estimated data from *Minerals Yearbook* 1974 and *Gas Facts* 1973. Constant 1970 dollars.

Figure 4-15. Investment data, United States petroleum industry

financed if the industry rate of return were high (corresponding to the recent total business external financing norm of 50 percent; see Hass, Mitchell, and Stone 1944, p. 108). Therefore, a maximum of 60 percent total revenues (2 x 30 percent) could be invested yearly if the incentive to invest (return on investment) were high enough.

The actual fraction of revenues invested in the COAL2 model varies with return on investment. This generally accepted measure of the financial health of an industry controls both the commitment of internal funds and the ability to attract external funds (Hass, Mitchell, and Stone 1974, pp. 22-25). The results of regression on historical petroleum industry data indicate a strong relationship between actual investment and the *average* return on investment of the industry (Figure 4-16).

The investment mechanism described above forms part of three feedback loops (shown in Figure 4-14). Under free market conditions a shortage of domestic production capacity requires companies to raise their profit margins. (Yet domestic producers could not raise domestic prices above foreign oil prices in an unregulated market. Thus foreign oil prices, not shown in Figure 4-14, provide a price ceiling for unregulated oil and gas prices in COAL2.) Increased profits tend to raise both revenues and return on investment, generating higher investments in oil and gas production facilities. After a construction delay, these new facilities become part of the total oil and gas production capacity. The price response is therefore part of two negative feedback loops (operating through revenues and return

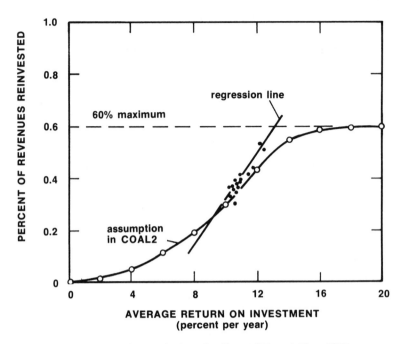

Source: Data from *Minerals Yearbook* 1974; *Gas Facts* 1973; and *Chase* 1974a.

Figure 4-16.   Oil and gas investment as a function of return on investment

on investment) that work to balance oil and gas supply and demand by adjusting new oil and gas investments.

When oil and gas prices are regulated, the link between the capacity/demand balance and profits is broken. Prices are no longer allowed to react to industry undercapacity, but are instead set to reflect an allowed rate of return on capital of 12 percent per year (Spritzer 1972, pp. 119-120). The deregulation of oil and gas prices and subsequent return to free market conditions (closing the two negative feedback loops in Figure 4-14) will be tested as a policy option in the last section of the chapter.

In addition to the price adjustment mechanism, changes in production of oil and gas also affect investments. For example, an increase in oil and gas production generates more revenues, increasing investment (all else being equal), and thus generating more production capacity. The resulting positive feedback loop creates a potential for continual growth in domestic oil and gas production capacity from reinvestment of revenues (revenue growth loop in Figure 4-14).

Oil and Gas Imports

From 1950 to 1975 the amount of oil and gas imported by the United States increased from 1.3 quads per year (0.6 mbpd) to about

13 quads per year (6 mbpd). Until 1973, world oil prices were a bargain relative to other forms of energy, averaging about 3 dollars per barrel (in 1970 dollars) delivered to Eastern U.S. ports. The postembargo price skyrocketed to about 12 dollars per barrel, a factor of 4 increase over previous prices.

As modeled in COAL2, both the price and the availability of imports have a profound impact on the behavior of the domestic oil and gas system. Unless interrupted by restrictive policies such as quotas or embargoes, it is assumed in the oil and gas sector equations that the difference between domestic demand and domestic production is imported. Import prices are assumed exogenous, set by foreign governments and domestic policy (tariffs). The COAL2 oil and gas sector allows testing of the long-term effects of such import restrictions as tariffs or quotas on the behavior of the oil and gas system.

OIL AND GAS SECTOR CAUSAL STRUCTURE

Figure 4-17 illustrates the feedback loop structure governing the production of oil and gas in the COAL2 model. Oil and gas can be

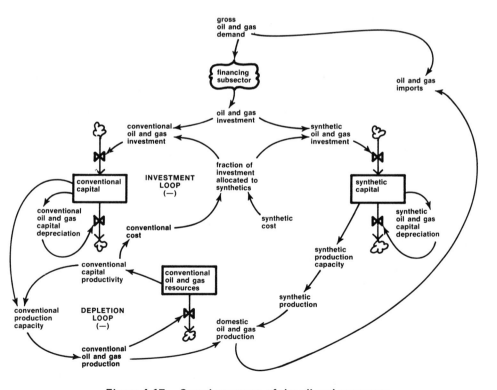

Figure 4-17. Causal structure of the oil and gas sector

produced from two sources: conventional wells and synthetic coal conversion facilities. The allocation of investment between synthetic and conventional production facilities is determined by a comparison of marginal production costs. Historically, virtually all oil and gas investments have been allocated to conventional sources, which have been available at about one-fourth the projected cost of synthetics. However, the growth in conventional oil and gas production has depleted the United States oil and gas resource base to the point where significantly fewer Btu's of oil and gas are currently produced per dollar of capital investment. As a result, conventional oil and gas production capacity has begun to decrease.

The depletion of oil and gas resources constitutes an integral part of two negative feedback loops. The negative "investment loop" (shown in Figure 4-17) attempts to minimize oil and gas production costs by allocating investment to the lowest-cost source of oil and gas. The "depletion loop" represents the physical effects of resource depletion on the efficiency (productivity) of capital. As oil and gas resources are depleted, cost increases cause investment to shift toward synthetic sources, and decreasing capital efficiency results in less oil and gas production per dollar of capital investment.

Before synthetic production facilities can be initiated, a significant amount of research and development (4-8 years) and construction (5 years) must be completed (FEA-STF 1974, p. 39). Such delays could prove to be an impediment to a smooth transition from conventional to synthetic sources of oil and gas. The next section examines the magnitude of the oil and gas transition problem and the relative effectiveness of various strategies designed to avoid a shortage of domestic oil and gas.

OIL AND GAS SECTOR SIMULATION RUNS

The severity of the United States energy problem can be measured by the difference between production and demand for oil and gas. If the United States is to avoid massive dependence on foreign oil, demand must shift away from liquid fuels and alternative sources of oil and gas must be developed to replace our dwindling resource base. The oil and gas sector is designed to determine what mix of three sources of oil and gas—conventional production, synthetic fuels, or imports—might satisfy demand over the next 35 years.

The simulation runs described in this section can be divided into four groups. First, a historical run compares the model-generated historical behavior with actual data. The historical simulation is then continued, as the reference run, to project the behavior of the United States oil and gas supply system to the year 2010 under a scenario of continued growth in demand. A number of policies are then tested to reduce the long-term dependence on oil imports projected in the reference run. Finally, various policy combinations are tested to design an effective long-term oil and gas strategy.

Historical Behavior

Figure 4-18 shows the behavior of the oil and gas sector during the historical period when driven with the historical values of gross oil and gas demand as an exogenous input. From 1950 to 1970 oil and gas production from conventional wells rises from 18.3 to 44 quads per year, equal to the real-world historical production rate. Production peaks in 1972 at 45 quads per year and falls thereafter to 42 quads per year in 1975. Because of the peak in production, oil imports rise rapidly after 1970 to meet the continually rising demand. Production from synthetic conversion of coal remains at zero during the historical period, for synthetics are uneconomical and commercially infeasible.

The exact timing of the peak in conventional oil and gas production depends on two factors: the amount of capital invested in new oil and gas production facilities and the effects of depletion on the oil and gas resource base. The historical return of 12 percent per year

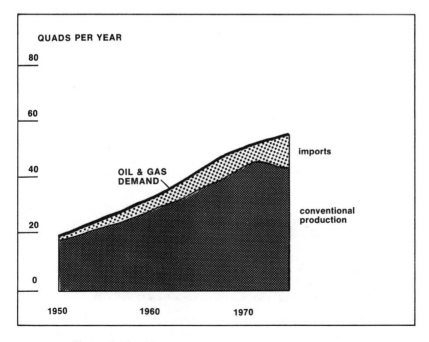

Figure 4-18. Historical behavior of the oil and gas sector

on domestic investment has not generated enough new domestic investment to meet demand. As a result, imports increase significantly from 1950 to 1970. Figure 4-18 also illustrates the initial effects of long-term oil and gas depletion, which exacerbates the capital investment problem. As oil and gas resources are nearly half depleted in 1970, capital costs begin to rise significantly, resulting in less output per dollar of capital investment. Regulated prices are unable to respond quickly enough to restore a balance between supply and demand. The peak and decline in domestic oil and gas production doubles United States dependence on foreign imports from 1970 to 1975.

Reference Projection of the Oil and Gas Sector

Figure 4-19 illustrates the behavior of the oil and gas sector if historical trends are allowed to continue. Gross oil and gas demand, an exogenous input to the sector, increases from about 55 quads per year in 1975 to 120 quads per year in the year 2010, a growth rate of about 2.2 percent per year (compared to a historical rate of almost 5 percent per year). This projection of demand is consistent with the reference projection of the demand sector of COAL2, where GNP increases at an average of 3.3 percent a year to the year 2000. Net oil and gas demand is assumed to slow its growth due to the shift

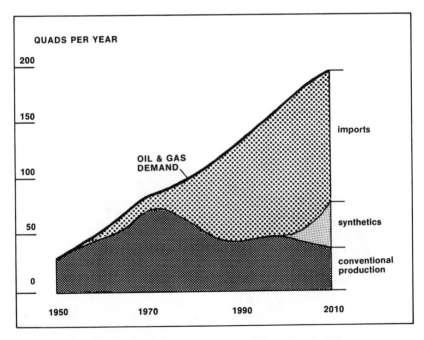

Figure 4-19. Reference projection of the oil and gas sector

to electricity over the long term. Considerable amounts of oil and gas are burned in utilities, due to sulfur restrictions on coal. Of the 100 quads per year of oil and gas demand in the year 2000 in the reference projection, 40 percent are burned in utilities to generate electricity.

Although oil and gas demand continues to rise past 1970 in Figure 4-19, conventional production from domestic wells peaks in 1970 and declines to less than one-half current levels by 1990. This decrease is caused by the combined effects of decreased productivity of new wells and limited domestic investment. The high capital cost of synthetic fuels prevents them from contributing significantly to supplies until after the year 2000. Production of oil and gas from synthetic conversion facilities reaches only 18 quads per year by the year 2010 (15 percent of demand). As a result, the reference run of the oil and gas sector projects a massive dependence on foreign oil imports throughout the transition period. The United States is dependent on imports for 60 percent of its oil and gas by 1985 and 70 percent by 2000.

The dependence on foreign sources depicted in the reference projection is likely to be politically intolerable. The 1985 import level depicted in Figure 4-19 is *four times* the stated import goal in the President's 1976 Energy Message (ERDA 1976, p. vii). Even if dependence on foreign sources for 70 percent of our energy were politically acceptable (as is currently the case in Europe), the import levels in the reference projection are high enough to seriously deplete the *foreign* oil and gas resource base by the year 2000, when combined with similar dependency projections for other countries. The reference or base case projection shown in Figure 4-19 therefore also represents a worst case projection, to be used to determine the effectiveness of various oil and gas policy options.

Oil and Gas Policy Analysis

A number of strategies have been suggested to achieve independence from foreign oil over the long term. The major suggestions include oil import restrictions, enhanced recovery techniques, price deregulation, accelerated synthetic programs, and oil and gas demand-reduction strategies. The effectiveness of each of these policies is tested in the following simulation runs by changing the model assumptions to represent the implementation of each policy in 1977.

Enhanced Recovery. One policy response to the problem of declining production of oil and gas is government support for the development of enhanced oil and gas recovery techniques. The ERDA plan (ERDA 1975, p. S-2) estimates that 40 billion barrels of oil (230 quads) and 250 trillion cubic feet of natural gas (260 quads) could be added to the recoverable oil and gas resource base by using enhanced recovery techniques, though at high capital costs. ERDA projects major energy contributions by 1985 (over 6 quads per year) and by the year 2000 (over 9 quads per year) from enhanced recovery (ERDA 1975, p. viii-1).

When the additional 500 quads of oil and gas from enhanced recovery are added to the total United States resource base of 2800 quads, the returns to drilling versus fraction remaining relationship changes, as shown in Figure 4-20. With the larger total resource base, replotting the data from Figure 4-6 shows that the capital productivity relationship shifts to the right of its reference value, since at each data point there is a larger fraction of resources remaining. Although enhanced recovery techniques have added to the total recoverable resource base, the additional resources are available only at or below current capital productivities (the region below 40 percent of resources remaining in Figure 4-20).

Figure 4-21 illustrates the effect of accelerated research, development, and demonstration of enhanced recovery technologies on the behavior of the oil and gas sector. Because the capital costs of the resources made available with enhanced recovery are quite high, the

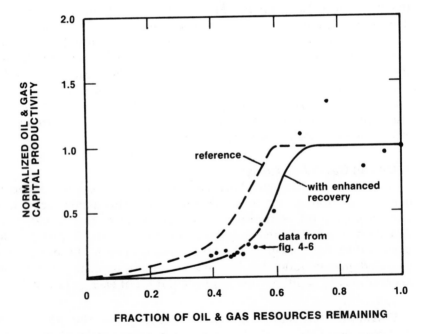

Source: Data from Figure 4-6 (Hubbert 1967).

Figure 4-20.   Enhanced recovery's effect on capital productivity

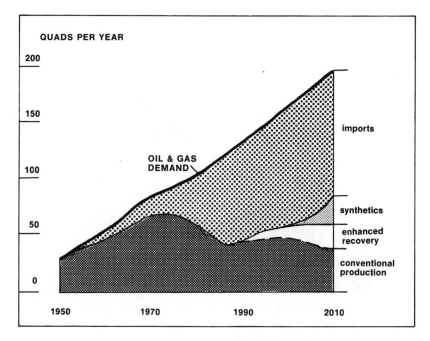

Figure 4-21. Enhanced recovery projection

enhanced recovery policy has little effect until after the year 1990. By the year 2010, conventional production is increased 12 quads per year (from 24 to 36 quads per year) with the use of tertiary recovery techniques.

Although the policy of government support to accelerate enhanced recovery techniques meets ERDA production goals, Figure 4-21 shows that enhanced recovery is ineffective when implemented as the sole policy solution. In this case, the oil and gas resources made available by tertiary recovery techniques are slow to be developed due to high costs and regulation-induced investment restrictions. The following run tests the effects of ameliorating the financial restrictions by deregulating the price of domestic oil and gas.

Price Deregulation. The reference projection assumes that oil and gas prices continue to be regulated. With price regulation, imports become the dominant source of oil and gas soon after 1975 (see Figure 4-19). Decontrol of oil and gas prices has been suggested as a potentially effective method of decreasing United States dependence on foreign oil. Figure 4-22 tests the supply effects of deregulating the price of domestic oil and gas in 1977. In this run deregulated prices adjust in an attempt to balance domestic demand with domestic

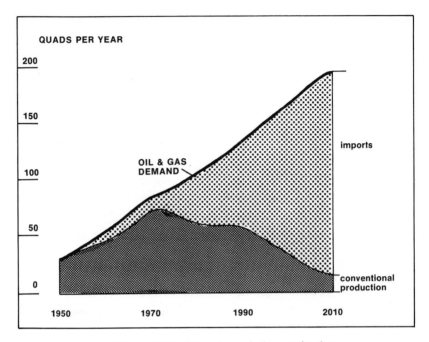

QUADS PER YEAR

Figure 4-22. Price deregulation projection

production: as demand exceeds production, prices increase to restore the balance. Yet domestic oil and gas prices cannot exceed foreign oil prices, for in a free market prices are determined by the cheapest marginal barrel of oil.

Figure 4-22 shows that price deregulation stimulates domestic production in the short run (to 1990) but could *increase* imports over the long run (to 2010). In the short run, deregulated prices are higher than in the reference run (which includes regulated prices), stimulating domestic production by as much as 30 percent in 1990. Consequently, accelerated depletion of oil and gas resources forces domestic costs to rise rapidly in the late 1970s and 1980s. Soon oil imports are the cheapest alternative (synthetic oil and gas must be priced at almost twice the 2 dollars per million Btu import price to be profitable), and the deregulated market sets oil and gas prices at the import price. Domestic investments are discouraged, causing domestic production to drop sharply after 1990 in Figure 4-22. By the year 2010 the United States imports 90 percent of its oil and gas.

Figure 4-22 suggests that one of the key questions for energy policymakers should be the design of pricing policies that ensure adequate investments in oil and gas during the transition period. As long as import prices remain above domestic prices, deregulation would stimulate domestic production (as occurs before 1990 in

Figure 4-22). Yet if imports are priced *below* domestic oil, price deregulation would not increase domestic oil and gas production over the long term.

Regulation at Higher Rates of Return. The return on invested capital of the United States petroleum and natural gas industries has averaged close to 12 percent per year from 1950 to 1972, slightly below the all-industry average of 12.3 percent per year in the same period (FEA 1976, p. 303; Chase 1974b, p. 14). But if the petroleum industry is to make a major shift to such high-risk, capital-intensive projects as coal-based synthetics, a higher-than-average return on investment will be necessary to generate sufficient new investment. Figure 4-23 illustrates the behavior of the oil and gas sector if oil and gas prices are regulated from 1977 onward to allow a 15 percent return on investment, 25 percent higher than the industry average.

By attracting and generating more funds for new investment, the oil and gas industry is able to increase domestic production by a growing amount over the long term. A policy that stimulates investment tends to have a cumulative long-term effect, for the yearly investment increases tend to accumulate fixed capital (refineries, drilling rigs, synthetic conversion plants). By the year 2000

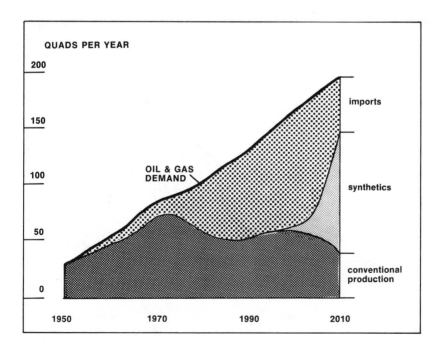

Figure 4-23. Regulation at higher rates of return projection

domestic oil and gas output is 25 percent higher than the base case (Figure 4-19). Past the year 2000 the additional investment accelerates the growth in synthetic fuels production to 60 quads per year in the year 2010. Imports are reduced to only 25 percent of oil and gas consumption by 2010.

Although a change in regulatory policy to increase the oil and gas industry's return on investment does reduce imports over the long term, the U.S. must still import over 60 percent of its oil and gas in the year 2000 (three times the 1975 dependency level). Part of the remaining problem lies with synthetic fuels: even though conventional oil and gas costs surpass synthetic fuel costs in 1990, production of synthetic fuels is delayed due to a lack of synthetic fuel R&D experience, and long lead times for plant design and construction. The next simulation tests a policy that anticipates the need for synthetic fuels by accelerating the research, development, demonstration, and commercialization of synthetic fuels.

Accelerated Synthetics Program. The production of conventional oil and gas peaked at 45 quads per year in 1972, and continues to decline in the reference run due to high capital costs of extraction. Even with such policies as enhanced recovery, deregulation, or

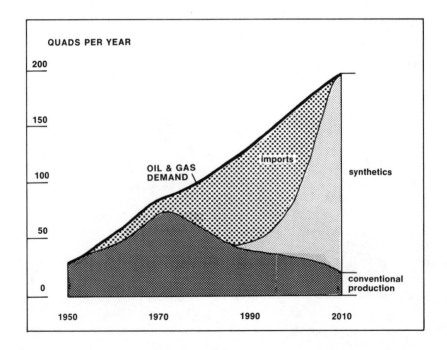

Figure 4-24. Accelerated synthetic fuels projection

regulation at higher prices, the production of oil and gas from conventional wells is unlikely to exceed early 1970 levels (see Figures 4-19, 4-21, 4-22, and 4-23). Given the limitations of conventional oil and gas production, an effective long-term policy would stimulate the development of synthetic fuels in anticipation of their eventual profitability.

The accelerated synthetic fuels program tested in this section consists of two parts. First, the research, development, and demonstration of synthetic fuel technologies is speeded up to allow construction of commercial synthetic facilities within 4 years of the start of the accelerated program, resulting in commercialization by 1981. Second, economic incentives, such as loan guarantees or a government-guaranteed price floor on synthetics, shift industry investment away from conventional oil and gas exploration and into the synthetic fuels program. By 1985, over 50 percent of all new oil and gas investment is channeled into the construction of synthetic conversion, processing, and transportation facilities.

Figure 4-24 shows the behavior of the oil and gas sector with an accelerated synthetic fuels program started in 1977. Although the program has little effect over the short run (to 1990), it is an attractive long-run option. Early commercialization of synthetic fuels provides more output per dollar of investment after 1990. Domestic oil and gas production is increased 80 percent in 2000 to 55 quads per year with the accelerated synthetics program. By the year 2010 synthetic fuels become the dominant source of oil and gas, satisfying over 90 percent of total demand. Imports decline from a peak of 50 quads per year in 1990 to zero in 2010 due to the increased production of synthetic fuels.

Demand Reduction Policies. Because it is so difficult to increase the supply of oil and gas over the short term, policymakers more frequently mention demand reduction as an immediate solution to the oil and gas problem. Final (net) demand for oil and gas could be reduced by implementing a number of new end-use technologies, described in the Ford Energy Policy Project "Technical Fix" scenario (Ford 1974b and Ford 1974a). In this scenario energy price increases encourage substantial use of such energy-conserving devices as heat pumps, better insulation, smaller and more efficient cars (25 miles per gallon), and more efficient steam production. With a concerted effort to reduce oil and gas consumption, the total (gross) demand for oil and gas could stabilize near current (1975) levels. In Figure 4-25 oil and gas demand stabilizes at 65 quads per year after 1990 (compared to 1975 consumption of 53 quads per year). By the year 2010 the demand reduction policy results in a 45 percent savings in oil and gas demand when compared to the reference projection.

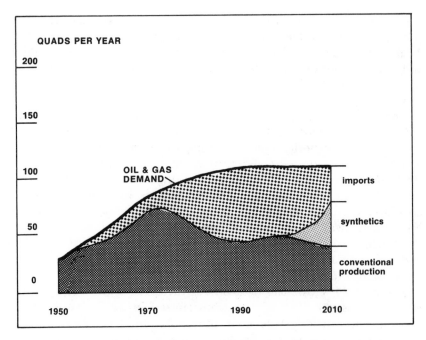

QUADS PER YEAR

OIL & GAS DEMAND

imports

synthetics

conventional production

Figure 4-25. Reduced demand projection

Figure 4-25 shows that conventional and synthetic oil and gas production levels are essentially unchanged from the reference projection, for, with price regulation, investment in domestic oil is not responsive to demand. Dependence on imports is still high from 1970 onward, even though demand for oil and gas has been stabilized over the long term. By the year 2000 the United States is dependent on foreign sources for over 50 percent of its oil and gas, compared to 70 percent dependence in the reference run.

Figure 4-25 illustrates that demand reduction policies alone are not sufficient to solve the long-term oil and gas problem. Because of the projected long-term decline in conventional oil and gas production, slowing the growth in demand is an insufficient policy response for the energy transition problem. A significant increase in domestic production (especially synthetic fuels) is needed if intolerably high dependence on foreign oil imports is to be avoided.

Combined Long-Term Policies. The previous simulation runs indicate that no single policy is capable of reducing United States dependence on foreign sources of oil and gas before the year 2000. Figure 4-26 shows the effects of combining the most effective policies: regulation at a higher rate of return (15 percent per year), demand reduction, and an accelerated synthetic fuels program.

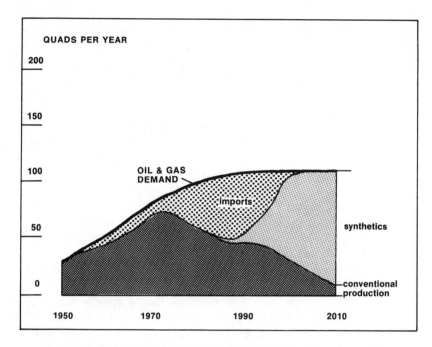

Figure 4-26. Combined policies projection for the oil and gas sector

As Figure 4-26 illustrates, these three policies tend to reinforce each other, thereby reducing imports to zero by the year 2000. Regulation of oil and gas prices at higher levels provides incentives to attract new capital for both conventional and synthetic production facilities, increasing domestic production significantly over the reference run (Figure 4-15). The accelerated synthetic fuels program anticipates the eventual depletion of conventional oil and gas resources, and shifts investment to synthetic fuels. As a result, production of synthetic oil and gas increases just in time as conventional production reaches the end of its life cycle (after 1990 in Figure 4-26). The above two policies combine with the reduced demand policies to cause imports to peak in 1988 at 35 quads per year (16 mbpd oil equivalent), and decline gradually to zero by 2000. Although independence from foreign oil is eventually achieved, the maximum dependence is still high, for imports supply over 50 percent of U.S. oil and gas in 1985—compared to 23 percent dependence in 1975.

CONCLUSIONS

Perhaps the most striking conclusion to be drawn from the simulations of the oil and gas sector of COAL2 is the sheer magnitude of the "liquid fuels" problem in the United States. The reference or base case behavior of the sector (Figure 4-19) projects a dramatic drop in domestic oil and gas production after 1975 and a rapid climb to 60 percent dependency on foreign oil in 1985. Furthermore, the problem persists, for imports supply 70 percent of U.S. oil and gas by the year 2000.

Deregulation of oil and gas prices is the policy most often suggested to achieve greater domestic production, and therefore reduce oil imports. Yet the analysis shown in Figure 4-22 suggests that deregulation stimulates domestic production only over the short term (to 1990), and the production increases are not dramatic (only 25 percent maximum increase in 1990). Over the long term, deregulation does not increase United States production of oil and gas, for oil imports become the least expensive supply option and dominate the domestic market after 1990.

All of the remaining policies tested do tend to increase oil and gas production over the long term, with varying degrees of effectiveness. Enhanced recovery policies are perhaps the least effective, stimulating oil and gas production by 6 quads per year by the year 2000 (Figure 4-21). The limited resource base and high capital costs of enhanced recovery techniques tend to limit their contribution. Policies that regulate oil and gas prices at higher levels, accelerate the commercialization of synthetic fuels, or reduce demand are all capable of more significant reductions in oil and gas imports (see Figures 4-23 through 4-25). In each case the United States must endure a 20-year period (1980-2000) where imports satisfy over 50 percent of United States oil and gas consumption.

The behavior of the oil and gas sector is greatly improved when policies are combined (Figure 4-26). Yet even if these policies were implemented in 1977, oil imports would continue to rise for another ten years to more than twice 1976 levels. Today's oil and gas problem has its roots in investment policies followed during the past 15-25 years. Furthermore, the momentum of past policies implies that the path of the oil and gas system is to a large extent already determined for at least the next ten years. This momentum is modeled in COAL2 as *delays*: time to perceive new trends, make decisions, and construct new production facilities. Current national energy policy goals still seek to reduce imports by 1985 (ERDA 1976, p.vii). To reverse the growth in imports from 1977 onward, policy changes were needed ten years ago—at least by 1967. There is no room for complacency in planning our energy future: the current oil import problems will escalate and persist over the long term unless policies such as those tested in Figure 4-26 are implemented immediately.

CHAPTER 4
EQUATIONS

```
100  *     OIL AND GAS SECTOR OF COAL2
110  NOTE
120  NOTE  OIL AND GAS SUPPLY-DEMAND BALANCE
130  NOTE
140  A     GOGD.K=NOGD.K+OGDU.K
170  A     DOGCUF.K=TABHL(DOGCUFT,GOGD.K/DOGPC.K,0,1.6,.2)
180  T     DOGCUFT=0/.2/.4/.6/.8/.85/.87/.89/.9
190  A     DOGPC.K=COGPC.K+SOGPC.K
200  A     OGCDR.K=OGC.K/GOGD.K
210  A     OGC.K=DOGPR.K+OGI.K
220  A     DOGPR.K=COGPR.K+SOGPR.K
230  NOTE
240  NOTE  OIL AND GAS FINANCING
250  NOTE
260  A     AOGP.K=(DOGP.K*DOGPR.K+OGIP.K*OGI.K)/OGC.K
270  A     DOGP.K=CLIP(UOGP.K,ROGP.K,TIME.K,DEREGT)
280  C     DEREGT=4000
290  A     ROGP.K=RPRAT.K*DCOST.K
300  A     RPRAT.K=CLIP(PPRAT,HPRAT,TIME.K,PYEAR)
310  C     HPRAT=1.86
320  C     PPRAT=1.86
330  A     UOGP.K=MIN(UPRAT.K*DCOST.K,OGIP.K)
340  A     UPRAT.K=TABHL(UPRATT,DOGPC.K/GOGD.K,0,2,.25)
350  T     UPRATT=6/5.4/4/3/2.4/1.86/1.5/1.2/1
360  A     DCOST.K=(CCOST.K*COGPR.K+SCOST.K*SOGPR.K)/DOGPR.K
370  A     DOGREV.K=DOGP.K*DOGPR.K
380  A     POGRR.K=CLIP(POGRR2.K,POGRR1.K,TIME.K,PYEAR)
390  A     POGRR1.K=TABHL(POGRR1T,AOGROI.K,0,.20,.02)
400  T     POGRR1T=0/.02/.05/.08/.13/.22/.37/.52/.57/.59/.6
410  A     POGRR2.K=TABHL(POGRR2T,AOGROI.K,0,.20,.02)
420  T     POGRR2T=0/.02/.05/.08/.13/.22/.37/.52/.57/.59/.6
430  A     AOGROI.K=SMOOTH(DOGROI.K,ROIAT)
440  C     ROIAT=5
450  A     DOGROI.K=(DOGP.K-DCOST.K)*DOGPR.K/(COGC.K+SOGC.K)
460  A     DOGINV.K=POGRR.K*DOGREV.K
470  NOTE
480  NOTE  CONVENTIONAL OIL AND GAS PRODUCTION
490  NOTE
500  L     COGC.K=COGC.J+(DT)(COGICR.JK-COGDR.JK)
510  N     COGC=COGCI
520  C     COGCI=45.8E9
530  R     COGDR.KL=COGC.K/ALOGC
540  C     ALOGC=20
550  R     COGICR.KL=DELAY3(COGIR.JK,COGCT)
560  C     COGCT=5
570  R     COGIR.KL=DOGINV.K*(1-FIASS.K)
580  A     CCOST.K=OGCCAF*(1/COGCCR.K)
590  C     OGCCAF=.175
600  A     COGCCR.K=CCCRI*OGCEM.K
610  C     CCCRI=.5E6
620  A     OGCEM.K=TABLE(OGCEMT,FOGRR.K,0,1,.1)
630  T     OGCEMT=0/.04/.09/.16/.26/.6/1/1/1/1/1
640  A     FOGRR.K=COGR.K/COGRI
650  C     COGRI=2.8E18
660  L     COGR.K=COGR.J+(DT)(-COGDPL.JK)
```

```
670 N        COGR=COGRI-COGP50
680 C        COGP50=.46E18
690 R        COGDPL.KL=COGPR.K
700 A        COGPR.K=COGPC.K*DOGCUF.K
710 A        COGPC.K=COGCCR.K*COGC.K
720 NOTE
730 NOTE     SYNTHETIC OIL AND GAS PRODUCTION
740 NOTE
750 L        SOGC.K=SOGC.J+(DT)(SOGICR.JK-SOGCDR.JK)
760 N        SOGC=SOGCI
770 C        SOGCI=0
780 R        SOGCDR.KL=SOGC.K/ALSYNC
790 C        ALSYNC=20
800 R        SOGICR.KL=DELAY3(SOGIR.JK,SCT)
810 C        SCT=5
820 R        SOGIR.KL=DOGINV.K*FIASS.K
830 A        FIASS.K=DLINF3(IFIASS.K,SDT)
840 C        SDT=8
850 A        IFIASS.K=CLIP(FIASS2.K,FIASS1.K,TIME.K,PYEAR)
860 A        FIASS1.K=TABHL(FIASS1T,CCOST.K/SCOST.K,0,2,.25)
870 T        FIASS1T=0/0/0/.1/.5/.8/.9/.95/1
880 A        FIASS2.K=TABHL(FIASS2T,CCOST.K/SCOST.K,0,2,.25)
890 T        FIASS2T=0/0/0/.1/.5/.8/.9/.95/1
900 A        SCOST.K=SPC.K+SFC.K
910 A        SPC.K=OGCCAF*(1/SYNCCR)
920 C        SYNCCR=.14E6
930 A        SFC.K=CPRICE.K/SCE
940 C        SCE=.6
950 A        SOGPC.K=SOGC.K*SYNCCR
960 A        SOGPR.K=SOGPC.K*DOGCUF.K*CPDR.K
970 A        CDS.K=SOGPC.K*DOGCUF.K/SCE
980 NOTE
990 NOTE     OIL AND GAS IMPORTS
1000 NOTE
1010 A        OGI.K=CLIP(ROGI.K,UOGI.K,TIME.K,PYEAR)
1020 A        UOGI.K=MAX(GOGD.K-DOGPR.K,0)
1030 A        ROGI.K=MIN(OIQ*GOGD.K,UOGI.K)
1040 C        OIQ=1
1050 A        OGIP.K=CLIP(FOP2.K,FOPH,TIME.K,EMYEAR)
1060 C        FOPH=.5E-6
1070 C        EMYEAR=1973
1080 A        FOP2.K=CLIP(FOPT,FOPE,TIME.K,TYEAR)
1090 C        FOPE=2E-6
1100 C        FOPT=4E-6
1110 C        TYEAR=4000
1120 NOTE
1130 NOTE     EXOGENOUS INPUTS
1140 NOTE
1150 A        NOGD.K=TABLE(NOGDT,TIME.K,1950,2010,10)*1E15
1160 T        NOGDT=18.3/30.4/45.5/50/56/60/66          •
1170 A        OGDU.K=TABLE(OGDUT,TIME.K,1950,2010,10)*1E15
1180 T        OGDUT=1.1/2.2/5.5/11/23/39/52
1190 A        CPDR.K=TABLE(CPDRT,TIME.K,1950,2010,10)
1200 T        CPDRT=1/1/1/1/1/1/1
1210 A        CPRICE.K=TABLE(CPRICET,TIME.K,1950,2010,10)*1E-6
1220 T        CPRICET=.35/.35/.35/.5/.6/.6/.6
1230 NOTE
1240 NOTE     CONTROL CARDS
1250 NOTE
1260 N        TIME=1950
1270 C        PYEAR=1977
1280 SPEC     DT=.2/LENGTH=0/PLTPER=2/PRTPER=0
1290 PLOT     GOGD=D,OGC=C,COGPR=O,DOGPR=P(0,1.2E1/)
```

```
1300 RUN
1310 NOTE
1320 NOTE    PARAMETER CHANGES FOR THE OIL & GAS SECTOR RUNS
1330 NOTE
1340 NOTE    HISTORICAL RUN
1350 NOTE
1360 C       PLTPER=1
1370 C       LENGTH=1975
1380 PLOT    OGC=C,COGPR=O,DOGPR=P(0,8E16)
1390 RUN     HISTORICAL RUN
1400 NOTE
1410 NOTE    REFERENCE RUN
1420 NOTE
1430 CP      PLTPER=2
1440 CP      LENGTH=2010
1450 PLOT    GOGD=D,OGC=C,COGPR=O,DOGPR=P(0,1.2E17)
1460 RUN     REFERENCE RUN
1470 NOTE
1480 NOTE    POLICY RUNS
1490 NOTE
1500 C       COGRI=3.3E18
1510 T       OGCEMT=0/.04/.08/.12/.18/.25/.7/1/1/1/1
1520 RUN     ENHANCED RECOVERY
1530 C       DEREGT=1977
1540 RUN     PRICE DEREGULATION
1550 C       PPRAT=2.07
1560 RUN     REGULATION AT 15 %/YR ROI
1570 C       SDT=4
1580 T       FIASS2T=0/.2/.5/.8/.9/.95/.98/.99/1
1590 RUN     ACCELERATED SYNTHETIC POLICY
1600 TP      NOGDT=18.3/30.4/45.5/50/55/55/55
1610 TP      OGDUT=1.1/2.2/5.5/11/11/11/11
1620 RUN     DEMAND REDUCTION POLICIES
1630 C       PPRAT=2.07
1640 C       SDT=4
1650 T       FIASS2T=0/.2/.5/.8/.9/.95/.98/.99/1
1660 RUN     COMBINED LONG-TERM POLICIES
```

※ *Chapter 5*

The Electricity Sector

PURPOSE OF THE ELECTRICITY SECTOR

The electricity sector of COAL2 simulates the electric utility industry's efforts to satisfy the long-term demand for electricity by constructing an adequate supply of new generating capacity. The model addresses two major issues critical to future development of the industry:

- financing new utility investments
- choosing the fuel mix of new electrical generating capacity

Since 1970, the electric utility industry has had increasing difficulty financing new capital investments. Increased costs combined with delays in obtaining price increases from regulatory boards have decreased the rate of return on utilities' investment, making it difficult for them to obtain new financing. Furthermore, uncertainties about future environmental legislation (SO_2), fuel availability, and capital costs have clouded the already-difficult decision about what type of new plant to build: nuclear, coal, or oil and gas.

The behavior of the electricity sector is important to the total energy system in two ways. First, although electricity comprised only 11 percent of net energy demand in 1975, its share is growing rapidly: electricity is expected to double its share of final demand by the year 2000 (ERDA 1975, pp. IV-4, V-3; USDI 1975, p. 36). Second, large efficiency losses magnify the role of electricity in the energy system, for each Btu of electricity output requires approximately 3 Btu's of fossil or uranium resource inputs. The demand for coal, oil, gas, and uranium will be significantly affected by the investment decisions of electric utilities during the next 30 years.

Because of the major shortages in domestic oil and gas supply projected in the oil and gas sector simulation runs (Chapter Four), the policy objective of the COAL2 electricity sector is to satisfy demand while reducing usage of oil and gas to a minimum. The effectiveness of the following policies is tested in the electricity sector simulation runs:

- changing regulatory procedures (rate relief)
- reducing daily and seasonal fluctuations in demand (load management policies)
- reducing electricity demand through conservation programs
- retrofitting coal-fired plants with stack-gas scrubbers (to reduce SO_2 emissions)
- relaxing SO_2 emissions standards
- adopting an accelerated nuclear program
- imposing a nuclear moratorium

HISTORICAL BEHAVIOR

The behavior of the United States electric utility industry over the past 25 years is characterized by two major phenomena:

- The industry has exhibited exponential growth in generating capacity and revenues at an annual rate of over 7 percent per year.
- The industry is currently experiencing a transition from a period of falling unit costs to a period of rising costs.

Figure 5-1 shows the historical growth in total U.S. electricity generation, which increased from 330 billion kilowatt-hours (1.1 quads per year electrical output converted at 3,412 Btu/kwh) in 1950 to 1,890 billion kilowatt-hours (6.4 quads per year) in 1975. As electricity generation has grown, so have total revenues from sales of electricity, Figure 5-2. Investments in new construction have varied more than generation and revenues, but have also grown significantly over the historical period (Figure 5-3). Cyclic fluctuations in investments are clearly evident in the investment data of Figure 5-3.

Over most of the history of the electric utility industry, unit costs and prices have been falling. Figure 5-4 shows the historical decline of fossil-fired steam-electric plant capital costs, representative of the capital costs for the bulk of the industry. Fuel costs (Figure 5-5)

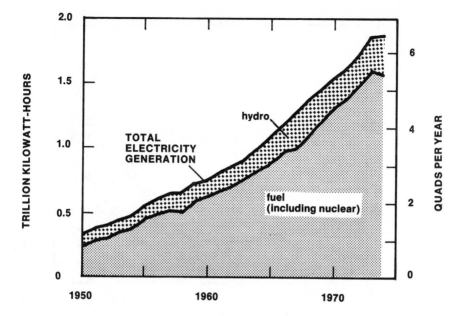

Source: EEI 1974a, p. 17. Used by permission.

Figure 5-1.   Total United States electricity generation

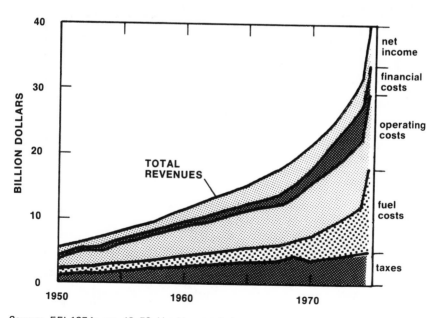

Source: EEI 1974a, pp. 42, 58. Used by permission.

Figure 5-2.   Electric utility revenues

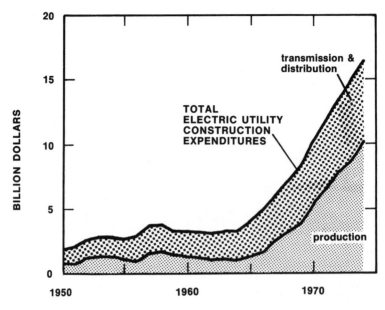

Source: EEI 1974a, p. 59.

Figure 5-3.   Investment in new plant and equipment by investor-owned utilities

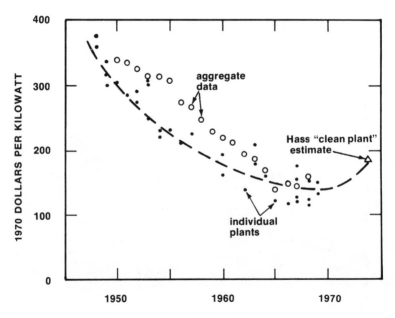

Source: Data from Hass, Mitchell and Stone 1974, p. 116.

Figure 5-4. Cost of capacity additions of fossil-fueled steam-electric plants

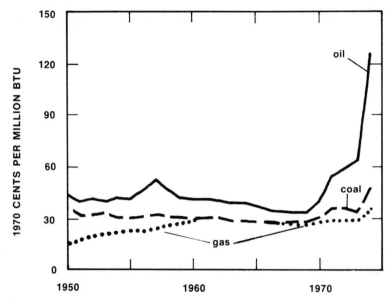

Source: Data from EEI 1974a, p. 50 (and earlier years).

Figure 5-5. Cost of fuels burned to generate electricity

have remained nearly constant throughout most of the historical period. As a result of these two trends, the average price paid by the electricity consumer (Figure 5-6) has declined in both current and constant dollar terms over most of this period.

Yet a number of recent trends indicate that the industry is entering a new period of rising costs. The decline in capital costs for new fossil and nuclear plants ended in 1965. Capital costs began to rise due to additional equipment needed for environmental protection, and increased delays in the construction of nuclear plants. Hass estimates the capital costs of a new fossil-fired facility with SO_2 removal equipment and some thermal pollution reduction apparatus to be 185 constant 1970 dollars per kilowatt in 1974 (Hass, Mitchell, and Stone 1974, p. 115), significantly above the trend line in Figure 5-4. Since the Arab Oil Embargo of 1973, fossil fuel prices have skyrocketed (see Figure 5-5), adding significantly to the unit costs of electricity. In the face of rapidly rising costs and inadequate rate increases, a number of utilities are currently facing financial difficulties, and new investments for 1974 and 1975 were significantly below historical trends.

Figure 5-7 shows historical trends in the mix of total electrical generation. Hydropower has grown more slowly than steam-electric generation, due to the gradual saturation of available sites for new

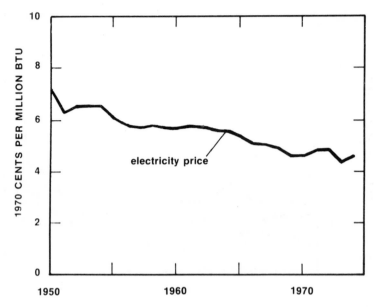

Source: EEI 1974a, p. 53. Used by permission.

Figure 5-6.   Average price of electricity to consumers

development. Thus its share of total generation has fallen since 1950 (Figure 5-7a). Of the fossil fuels consumed for electricity generation, Figure 5-7b shows that coal's share has declined somewhat since 1965, and has been largely replaced by oil in order to meet environmental standards.

Figure 5-8 shows the trends in new generating capacity orders, representative of the allocation of new utility investment by fuel type. Nuclear's share of new plant orders has increased from no orders in 1960 to around 40 percent (Figure 5-8a), as investment in nuclear power became both technologically feasible and economically competitive. Coal's share of new fossil-fired boiler orders dropped rapidly during the 1960s in response to SO_2 legislation. In the past five years, however, coal's share of new fossil-fired orders has begun an upswing in response to the rapid rises in oil costs since 1973 (Figure 5-8b). Future behavior of each fuel's share of new investment will depend on fuel cost trends, environmental legislation, and the maturation of new technologies such as SO_2 scrubbers.

BASIC CONCEPTS

Utility Financing: Capacity Planning

In the COAL2 model, electricity *generation* represents the amount of electricity actually generated in a given year to satisfy annual

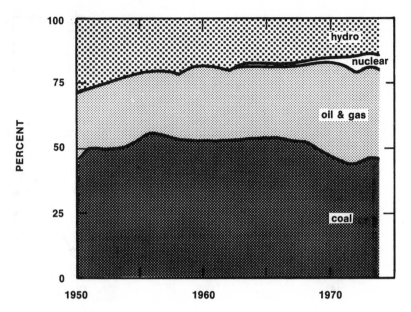

Source: EEI 1974a, p. 18.

a. Share of total generation

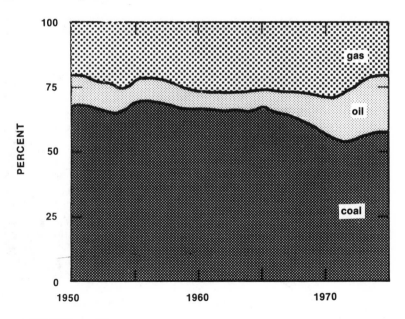

Source: EEI 1974a, p. 22.

b. Share of fossil generation

Figure 5-7.   Historical mix of electrical generation

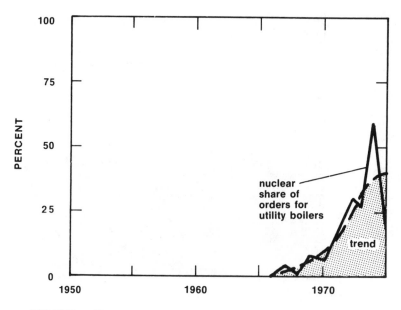

Source: AEC 1974, p. 12.

a. Nuclear's share of new steam-electric capacity orders

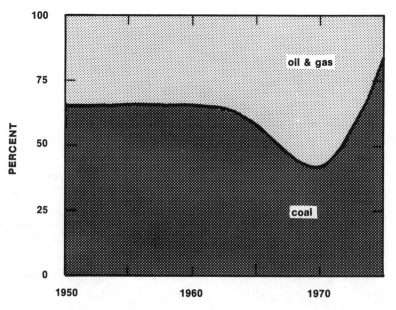

Source: Steam-Electric Plant Factors 1975, p. 56 (and earlier years).

b. Coal's share of new fossil-fired capacity orders.

Figure 5-8.   Historical trends in new generating capacity orders

demand, and is measured in kilowatt-hours/year (or quads/year). *Capacity* measures the plant's ability to generate electricity, measured in kilowatts (or quads/year). In addition to its long-term growth trend, the demand for electricity has exhibited daily, weekly, and seasonal fluctuations of considerable magnitude. Because demand fluctuates, the capacity required to satisfy peak loads (plus some margin of safety, normally 20 percent) is significantly greater than average generation rates. The annual capacity utilization factor (electricity generation ÷ rated capacity) has remained near 50-55 percent over most of the history of the electric utility industry (EEI 1974a, pp. 5, 18).

The basic mechanisms controlling the planning and financing of new generating capacity in COAL2 are shown in Figure 5-9. The capacity planning loop structure attempts to balance electricity demand and capacity by adjusting investment in new generating capacity. When the capacity factor exceeds its normal value of 55 percent, reserve margins are low, and investment increases. Similarly, deferrals of planned construction can be attributed to a drop in the capacity factor. In 1975, the capacity factor dropped to 44 percent, the reserve margin increased to 37 percent, and a total of 215,100 megawatts of new capacity orders were deferred or cancelled (FEA 1976, pp. 232, 237). If electricity demand again begins to grow at its expected rate of 5 percent per year or greater after 1975, the capacity factor would increase, calling for greater investment in new capacity.

Building new capacity also creates a larger revenue base from which investment can be drawn. When these new investments add to generating capacity, a positive feedback loop is closed (revenues growth loop). This second feedback loop allows the industry to grow by its own momentum, until limited by demand through the capacity planning loop.

Utility Financing: Regulation

Before the 1973 oil embargo and recession, financing new utility capacity was a straightforward process. Demand was predictable, increasing at a steady 7 percent per year. Furthermore, technological advances of the post-World War II period brought improved efficiencies in generating equipment, and decreased generating costs. Because public utility commissions set rates based on historical costs, utilities typically earned more than the allowed rate of return targeted by the commissions. Since utility securities and bonds were considered one of the safest of investments, utilities generally had no trouble financing new construction during this period.

In the mid-1960s, however, the long trend of improved efficiency in generating electricity came to a halt. Because of increased concern

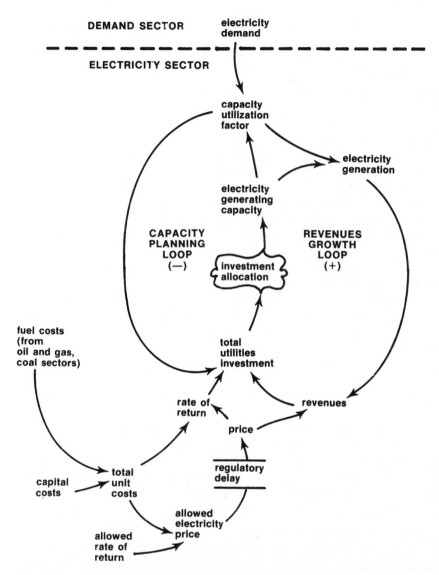

Figure 5-9. Financial structure of the electricity sector

for the environment, additional equipment for environmental protection became part of the capital base for any new fossil-fired plant, and capital costs began to rise. The fuel cost rises of 1973 further exacerbated the rise in unit costs. Because of the delays in obtaining rate relief through the regulatory process, the utilities' financial position is undermined as costs rise. In the early 1970s, industry earnings were on the decline, and utility bonds were derated (Hass,

Mitchell, and Stone 1974, p. 65). In 1974 and 1975, new investments fell well below historical levels as utilities experienced difficulty selling new bond issues (BW 1975).

The basic mechanisms of the regulatory structure included in COAL2 are sketched in Figure 5-9. (A more detailed model of the financial structure of the U.S. electric utility industry is the subject of a Dartmouth substudy: Ford, F.A. 1975.) The allowed electricity price is calculated from current unit costs and the allowed rate of return. Because it takes a significant amount of time for regulatory commissions to perceive and measure cost changes, hold hearings, and approve rate changes, the current price of electricity is rarely exactly set at the hypothetical allowed electricity price. Due to regulatory delays, the actual price lags the allowed electricity price by an average of two years (Secrest and Burzlaff 1974, p. 16).

When costs begin to rise, as they did in the late 1960s, the regulatory lag causes the industry's rate of return to drop. In the COAL2 structure (Figure 5-9), if the rate of return remains persistently low, the industry's ability to finance new planned capacity (as determined by the capacity planning loop) is imperiled. Policies designed to improve the financial health of utilities can be tested in COAL2 in two ways: by increasing the allowed rate of return, or by reducing the regulatory lag.

The Utility Fuel Mix Decision

As the utilities raise new capital (the financing decision), they must also decide how to allocate it among alternative methods of generating electric power. Because the potential for growth in hydropower is limited, the investment decision focuses on the choice of boiler fuel for steam-electric plants: uranium, coal, or oil and gas. The utility fuel-mix decision is a critical part of the U.S. energy problem, for the choice of boiler fuels could either add to or help alleviate the oil and gas shortage. Unless adequate nuclear and coal-fired generating capacity can be built, future increases in electricity demand must be met by burning additional amounts of our scarcest fuels, oil and gas.

Figure 5-10 illustrates the fuel-mix decision structure included in the COAL2 model. It is assumed that with no environmental constraints, investment is allocated to the least-cost fuel option. Total investment (obtained from the utility financing sector) is first divided between fossil and nuclear facilities, based on a comparison of the unit costs of fossil and nuclear electricity. Investment in fossil-fired utilities is further divided between coal-fired and oil- and gas-fired utilities, based on a comparison of fuel costs. If SO_2 emissions increase to the point where federal standards are violated,

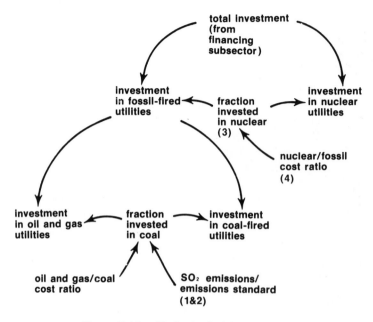

Figure 5-10. Fuel mix decision structure

utilities are forced to cease constructing the less-costly coal-fired utilities and build oil- and gas-fired units which meet federal emission standards.

The fuel-mix structure illustrated in Figure 5-10 allows one to test the following policies:

1. installation of stack gas scrubbers
2. relaxation of SO_2 emissions standards
3. nuclear moratorium
4. reduced nuclear costs and siting delays

The first three policy options accelerate the use of coal for generating electricity, while the fourth policy accelerates the use of nuclear power. The structural element affected by each numbered policy is indicated by matching numbers in Figure 5-10. Each of these policies is tested in runs later in this chapter.

ELECTRICITY SECTOR CAUSAL STRUCTURE

The causal structure of the electricity sector is illustrated in Figure 5-11. Generating electricity is actually a conversion process, and is so represented in COAL2. Three different conversion process capital stocks are represented explicitly in the structure: nuclear utilities,

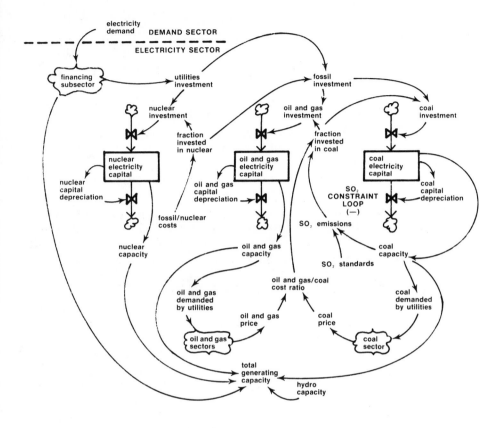

Figure 5-11. Causal structure of the electricity sector

coal-fired utilities, and oil- and gas-fired utilities (hydropower generation, treated exogenously, is not shown in Figure 5-11). Resources from the oil and gas or coal sectors are converted to electricity by these facilities and delivered to the final consumers of energy in the demand sector. Therefore, no resource stocks are represented in the electricity sector—the fossil-fired utilities within the sector must draw on the finite resource stocks of the oil and gas or coal sectors.

Because producing electricity is a capital-intensive process, the focus of the structure is on financing—raising funds and allocating them among nuclear, coal-fired, and oil- and gas-fired utilities. The financing and fuel decisions control both the total amount and type of generating capacity available at any point in time.

The electricity sector contains a major environmental feedback loop, the SO_2 constraint loop shown in Figure 5-11. New capital investment, raised in the utility financing subsector, is normally divided between nuclear, coal-fired, or oil and gas-fired capacity according to a comparison of costs. Yet, if SO_2 levels begin to violate federal emissions standards, investment in coal-fired utilities is decreased. The link between coal-fired capacity, SO_2 emissions, and the decision to invest in new capacity forms a negative feedback loop that constrains investment in coal-fired utilities based on environmental considerations.

A number of other feedback loops affecting the fuel-mix decision are closed when the electricity sector is linked to the other sectors of COAL2. Oil and gas prices and coal prices respond to the utilities' demand for oil and gas or coal. For example, a large shift to coal as a boiler fuel would increase coal prices, which would make an investment in coal-fired utilities a less attractive option. If the coal price hike were large enough, coal-fired capacity would decrease over the long term, completing a negative feedback loop through the coal sector. A similar price-sensitive feedback loop is closed through the oil and gas sector of COAL2. These two feedback loops tend to offset any cost advantage that may develop for a given boiler fuel.

ELECTRICITY SECTOR SIMULATION RUNS

This section examines the future behavior of the electricity sector for potential capacity shortages and fuel mix problems. First, the sector's behavior is compared with real-world historical behavior as a check on model validity. The electricity sector of COAL2 is then run separately with exogenous forecasts of total electricity demand, oil and gas prices, and coal prices. The first series of projections test the vulnerability of the electric utility industry to capacity shortages, and the need for policy changes (rate relief or load management) to avoid a future shortage. A number of policies designed to avoid new construction of oil- and gas-fired utilities (an accelerated coal program, relaxation of SO_2 emission standards, or implementation of SO_2 emission control devices) are tested in a second series of projections.

Finally, a combination of financial and fuel-mix policies are examined in an attempt to design an effective long-term utility management program. To determine the feasibility of meeting future electricity demand with reduced dependence on nuclear power, a nuclear moratorium policy is also tested. In Chapter Seven, the need for these policies will again be examined when the COAL2 model is run in its entirety to simulate the total energy system.

Historical Behavior

Figure 5-12 illustrates the historical behavior of the COAL2 electricity sector from 1950 to 1975, when historical values for total electricity demand, oil and gas prices, and coal prices are used as

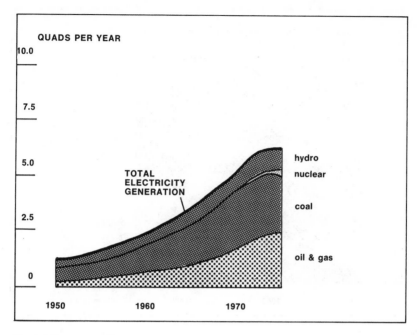

a. Electricity generation

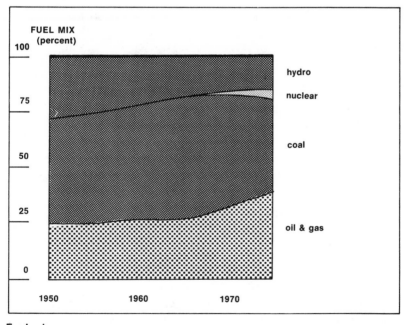

b. Fuel mix

Figure 5-12.   Electricity sector historical behavior

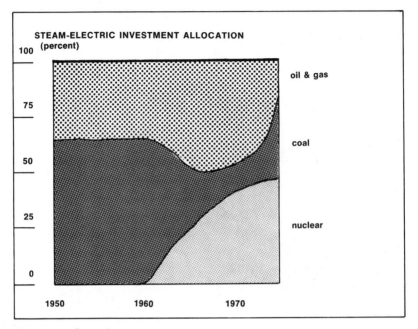

c. Allocation of new investment

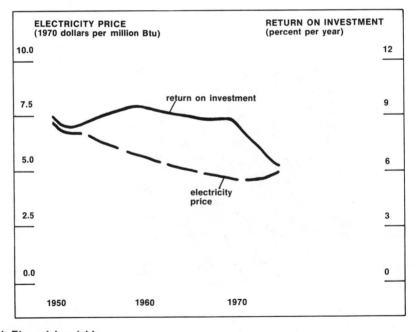

d. Financial variables

Figure 5-12 (continued).   Electricity sector historical behavior

exogenous inputs. Total electricity demand grows at over 7 percent per year from 1950 to 1970 (Figure 5-12a), but slows its growth rate considerably from 1970 to 1975 due to the recession. Electricity demand is equal to 6.4 quads per year (1.9 trillion kwh) in 1975.

The mix of fuels that satisfied electricity demand during the historical period is shown in Figure 5-12b. Hydropower generation increases more slowly than steam-electric generation during the historical period, resulting in a drop of its share of total generation from 30 percent in 1950 to 16 percent in 1975. A rapid increase in nuclear's share of new investment (Figure 5-12c) has caused nuclear power generation to grow from negligible amounts in 1960 to about 7 percent of total generation in 1975.

Until 1960, the fraction of total electricity generated from coal had remained relatively stable at 50 percent. After 1960, investment in new coal-fired utilities began to decline as a result of environmental legislation. Figure 5-12b shows that the fraction of electricity generated from oil and gas has increased rapidly during the late 1960s to over one-third of total electricity generation. The rapid rise in oil prices after 1973 combined with the use of low-sulfur coal stimulated an increase in coal-fired utilities investment in the early 1970s (see Figure 5-12c). Yet, at the end of the historical period, generation of electricity from oil- and gas-fired utilities is still increasing (Figure 5-12a), further worsening the domestic oil and gas shortage.

As shown in Figure 5-12d, electricity prices dropped from 1950 to 1970 as a result of the decline in unit costs of electricity. Since unit costs were declining, utilities typically earned more than the allowed rate of return of 8 percent per year (the COAL2 model output shows a 9 percent per year return on investment from 1950 to 1970 in Figure 5-12, consistent with historical data). The same regulatory lag turned against the companies, however, when costs began to rise in the late 1960s. In Figure 5-12d, return on investment began to decline after 1970 as the utilities' profits declined. The potential vulnerability of the electric utility industry to future capacity shortages and the need for rate relief or load management policies are examined in the following projections.

Reference Run

Figure 5-13 shows the reference projection of the COAL2 electricity sector. Electricity demand is assumed to grow at 5.4 percent per year from 1975 onward, equal to the FEA reference projection (FEA 1976, p. 238), but a significant reduction from the historical growth rate of over 7 percent per year. The assumed decrease in demand growth is caused by the decline in the growth rate of total net energy

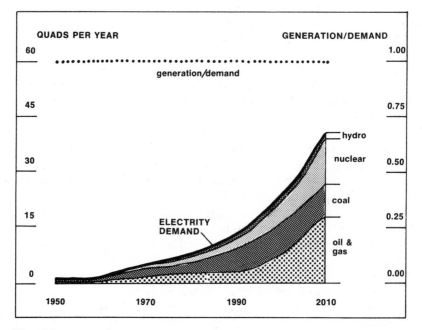

a. Electricity generation

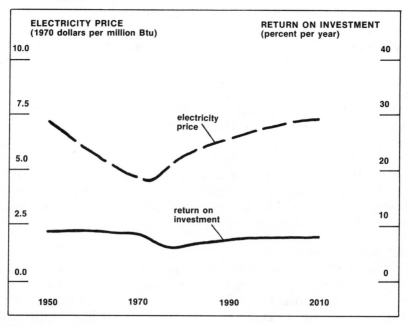

b. Financial variables

Figure 5-13.   Electricity sector reference projection

demand (2.2 percent per year instead of the historical 3.2 percent per year). Also included in the projection shown in Figure 5-13 is the assumption that oil and gas boiler fuel prices triple by the year 2000 (in constant dollars), and coal prices increase by 70 percent.

Figure 5-13 indicates that an electricity demand growth rate of 5.4 percent per year is not high enough to generate significant capacity shortages over the long term, though the utilities remain precariously vulnerable during the 1980s. Even with current regulatory practices, the timely decrease in demand growth allows generation to just barely keep up with demand over the long term. The adequacy of generating capacity is measured by the generation-demand ratio in Figure 5-13a. A generation-demand ratio less than 1.0 indicates the occurrence of a shortage or brownout in the COAL2 model. In the reference run, the financially hampered utilities begin to recover slightly after 1980, increasing their return on investment as costs begin to stabilize. Utilities manage to raise enough new capital to avoid any shortage of generating capacity over the long term.

Figure 5-13a also illustrates the projected electricity fuel mix to the year 2010. Although nuclear power grows at over 10 percent per year from 1975 to 2010, nuclear power satisfies only 30 percent of total electricity demand by the final year. The nuclear projections in COAL2 are well below the early AEC studies (AEC 1972, p. 4; AEC 1974, p. 2; FEA-NTF 1974, p. 3.3.5), but are nearer to current projections, which include the latest series of deferrals and cancellations (FEA 1976, p. 36). Nuclear power does not grow as in earlier projections in COAL2 due to the capital constraints and construction delays incorporated in the model. By 1975, the costs of electricity from nuclear and fossil-fired plants have reached parity, as the capital cost advantage of fossil-fired plants is offset by increased fuel prices. After 1975, investment in fossil and nuclear plants is divided almost equally in COAL2. Yet even with such a large commitment to nuclear power, fossil-fired generation remains the dominant source of electricity for the United States through the transition period.

Because the reference projection assumes that utilities begin to switch to low-sulfur fuel in 1970, electricity generated from coal grows substantially from 1970 to 1990 in Figure 5-13a. Yet the trend to coal-fired utilities is short-lived. After 1990, electricity generated from coal has again grown to the point where SO_2 emission standards are threatened—even with the use of low-sulfur coal. As occurred in 1960, investments in coal utilities are curtailed after 1990 due to SO_2 restrictions, causing electricity generated from oil and gas to grow to almost 50 percent of total electricity by 2010.

The Financial Vulnerability of Utilities
Even though the electricity sector reference run does not project

any shortages of generating capacity, the utilities come precariously close to brownouts in the 1980s due to their weak financial situation. The following two simulations test the vulnerability of the utilities' current financial situation, and the effectiveness of rate relief and load management programs to provide financial relief.

High Demand Growth Projection. Electricity demand is forecast to grow at 5.4 percent per year in the reference run of COAL2. Yet this projection is far from precise—demand growth could vary by as much as 2 percent per year due to uncertain assumptions or policy changes (such as oil and gas price deregulation). Figure 5-14 illustrates the behavior of the electricity sector if demand should return to its historical growth rate of 7.5 percent per year from 1976 to the year 1990.

Figure 5-14a shows that if total electricity demand were to resume its historical growth rate, generating capacity shortages would develop after the year 1980. These shortages appear as a drop in the electricity generation-demand ratio, which must equal 1.0 to avoid brownouts. With historical demand growth rates, 2.5 percent of annual electricity demand would not be met in 1988, the worst year of capacity shortages. The cause of the shortages is the gradual erosion of the utilities' financial position, shown as a decline in return on investment in Figure 5-14b.

During the recession of 1974-1975, electricity demand ceased to grow, remaining near its 1973 level of 6.3 quads per year (1,850 billion kwh). By the end of 1975, peak capacity had increased to 34 percent over that summer's peak demand, well above the acceptable margin of 20 percent (FEA 1976, p. 234). This excess reserve margin provides a cushion which will help avoid brownouts as demand begins to grow in the future. In spite of this, if demand returns to its historical growth rate of 7.5 percent per year, Figure 5-14 indicates that there will be brownouts over the long term unless the utilities obtain financial relief.

Rate Relief and Load Management. If the utility rate adjustment procedures were restructured, utilities could adjust more quickly to the rising costs of new generating plants. The lag between cost changes and rate adjustments (regulatory lag) could be minimized by one or more of the following procedural changes (Joskow 1974, p. 39):

- automatic cost adjustments
- temporary rate increases
- use of projected or "future-test-year" cost data

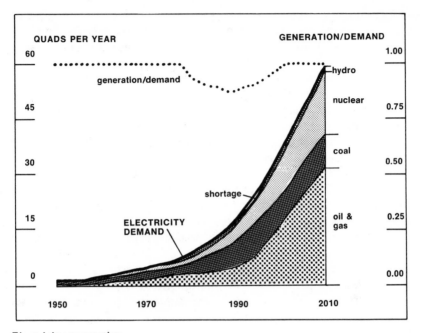

a. Electricity generation

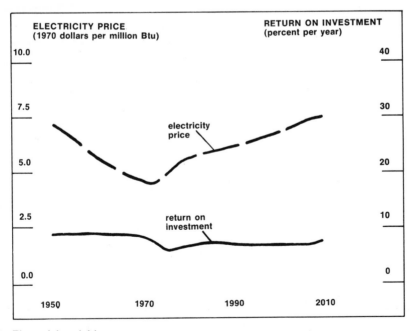

b. Financial variables

Figure 5-14. High demand growth projection

Rate commissions might also increase the allowed rate of return to reflect the increased risk involved in today's utility investments.

Utilities could also obtain some financial relief through a variety of load management programs. Such programs improve the "shape" of the load curve representing the daily and seasonal fluctuations in electricity demand. Thus load management increases the utilization of capacity and reduces future capacity expansion requirements. FEA estimates that through a combination of pricing and load controls, the annual capacity factor could be increased from its 1975 low of 44 percent to 57 percent in 1985 (FEA 1976, p. 237).

Several parameters were changed in the electricity sector model to simulate the effects of rate relief and load management. To represent rate relief, the allowed rate of return on total capital was increased from 8 to 10 percent per year. To model streamlined regulatory procedures and the use of future test-year cost data in the rate calculations, the structural lag between cost changes and rate adjustments was eliminated. To simulate load management, the normal capacity factor was increased to 60 percent. A higher capacity factor affects both the financing and the fuel-mix decisions: financial requirements are reduced, and the unit costs of generation for each fuel are changed as capital costs are spread over more output units.

Figure 5-15 illustrates the effects of rate relief and load management on the behavior of the utility industry. Electricity demand is assumed to grow at its historical rate (the same as in Figure 5-14), high enough to create a capacity shortage without these two policies. Yet rate relief and load management in 1977 successfully avoid the capacity shortage of the 1980s projected in the previous run. Rate relief policies increase the price of electricity to the point where producers earn their allowed rate of return of 10 percent per year. Figure 5-15b shows that this is accomplished with less than a 10 percent increase in electricity prices. The higher rate of return facilitates the utilities' acquisition of new capital over the long term. When combined with the load management policy, the reduced financial needs and increased financing capacity ensure that even a 7.5 percent per year growth rate in electricity demand could be satisfied.

Fuel Mix Policies

Although a combination of rate relief and load management policies tends to protect the utility industry against a shortage in generating capacity, the simulation shown in Figure 5-15 demonstrates that such policies have little effect on the fuel mix. The amount of electricity generated from oil and gas continues to grow in all three simulations (Figures 5-13 to 5-15), adding to the oil import

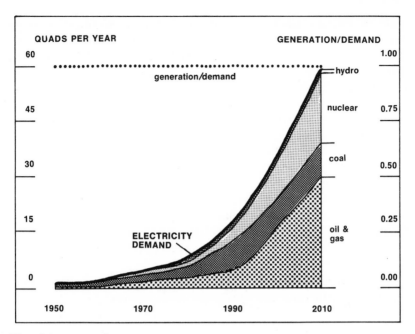

a. Electricity generation

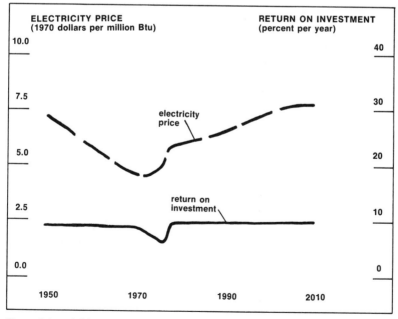

b. Financial variables

Figure 5-15.   Rate relief and load management

problem. Oil and gas burned in utilities grow from about 7 quads per year in 1975 to 10 quads per year in 1985 and 40 quads per year in 2010 in the reference projection. The added demand for oil and gas from utilities would seriously increase oil imports above the current 1975 level of 13 quads per year.

The following set of simulations tests the effectiveness of policies designed to avoid continued growth in the use of oil and gas as a utility boiler fuel by shifting the pattern of investment in new utilities away from oil and gas fired plants. First, a program designed to accelerate the growth in nuclear power is tested. Investment in coal-fired utilities is then encouraged by releasing the environmental constraints to coal use through both technological policies (SO_2 scrubbers) and legislative changes (relaxation of SO_2 standards). Finally, a combination of policies is suggested as the best method to reduce oil and gas use while satisfying electricity demand.

Accelerated Nuclear Program. The Atomic Energy Commission publishes yearly forecasts of nuclear power growth, based on "business-as-usual" policies and policy changes that tend to increase nuclear power plant capacity additions (AEC 1974, p. 5). To simulate an accelerated nuclear policy, planning and construction delays are reduced from 10 years to 6 years in COAL2, as a consequence of the following policy changes (AEC 1974, p. 6):

- standardizing plant designs to streamline the license application
- permitting construction to begin prior to completion of the construction permit application safety review
- completely separating the site environmental review from the safety review

In addition, nuclear fuel prices are reduced 20 percent by 1990 (and more thereafter) in the accelerated nuclear projection due to increased federal subsidies for the nuclear fuel cycle program. To secure the necessary uranium reserves, a detailed program to accelerate discovery and conversion of the nation's remaining uranium resources is coordinated between the U.S. Geological Survey, ERDA, and the Department of the Interior. As suggested by a recent National Academy of Sciences panel on uranium availability, such a program might include expanded geological mapping and exploration, research and development of new exploration techniques, and expediting property acquisition in the public domain (COMRATE 1975b, pp. 5-6).

The combined effects of the decreased nuclear construction delays and the decreased nuclear fuel price causes nuclear capacity to

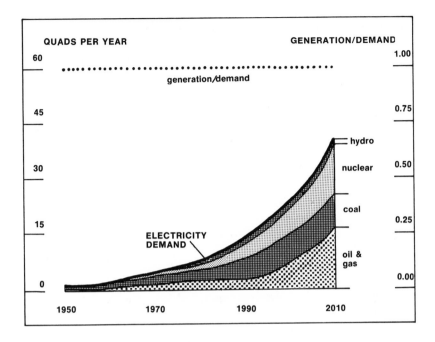

QUADS PER YEAR   GENERATION/DEMAND

generation/demand

hydro

nuclear

coal

ELECTRICITY
DEMAND

oil &
gas

1950   1970   1990   2010

Figure 5-16. Accelerated nuclear projection

increase to about 140 gigawatts by 1985 (Figure 5-16), slightly lower than the FEA accelerated nuclear projection of 168 gigawatts (FEA 1976, p. 38). Nuclear power increases its share of total electricity from 28 percent in the year 2010 in the reference run (Figure 5-13) to almost 35 percent. Yet because electricity generated from coal is constrained by environmental effects, oil and gas must still provide the bulk of the electricity generated from fossil fuels. With the accelerated nuclear program, oil and gas electricity generation still grows significantly, increasing to 7 quads per year in the year 2000. Generating 7 quads per year of electricity requires almost 20 quads per year (9 mbpd) of oil and gas for utility boiler fuel. Although an accelerated nuclear program certainly can help reduce oil and gas demand, it is by no means capable of solving the utility fuel mix problem.

Accelerated Coal Program—Relaxation of SO_2 Standards. A policy often suggested to increase electricity generated from coal is relaxation of current SO_2 emission standards. Any policy that allows increased coal use without reducing SO_2 emissions is considered a relaxation of SO_2 emission standards in this analysis. A large number of policies currently under consideration for legislation fall in this category:

- amend the Clean Air Act to allow more emissions
- postpone the deadlines for compliance with the Clean Air Act
- allow the use of tall stacks, rural siting, or intermittent controls to satisfy ambient standards

The controversy over methods that relax standards versus methods that reduce emissions (low sulfur coal, stack gas scrubbers) has been heated. While some utilities vehemently contend that SO_2 scrubbers are not reliable (Cook 1974, p. 66), both the Environmental Protection Agency and a National Research Council study conclude that they are ready for commercialization (EPA 1974, pp. 10-12; CNR 1975, pp. xxx-xxxi). In order to test the effects of relaxing SO_2 emission standards, allowed SO_2 emissions (measured in tons of SO_2 emitted per year) are increased 50 percent after 1977 in the simulation shown in Figure 5-17.

When SO_2 emission standards are relaxed, almost all of new fossil-fired utility construction is built to burn coal. After a 5-year construction delay, electricity generated by coal-fired plants grows significantly, causing oil- and gas-fired generation to drop in the 1980s. By 1990, coal satisfies 50 percent of total electricity demand, and oil and gas is required for only 20 percent (compared to 33 percent satisfied by oil and gas in 1975).

Yet the growth in coal-fired generating capacity lasts only 15 years. In that time, emissions grow to violate the relaxed SO_2 standards, even though low-sulfur coal is being used extensively. SO_2 restrictions curtail investments in coal-fired utilities after 1990, forcing renewed construction of oil and gas-fired power plants. By the year 2010, electricity generated from oil and gas grows to 16 quads per year, demanding 45 quads per year of oil and gas as a boiler fuel. Oil- and gas-fired utilities generate 40 percent of the total by the year 2010, while coal's share of the electricity market declines from 50 percent in 1990 to 30 percent by the year 2010.

Figure 5-17 demonstrates that a relaxation of SO_2 emission standards for coal-fired plants is a shortsighted policy, resulting in a short spurt of growth in coal-fired capacity followed by a renewed period of environmental constraints. The problem of dependence on oil and gas as a utility fuel reappears shortly after the new set of emission constraints are imposed. Furthermore, during the 15 year period of growth in coal-fired capacity, SO_2 emissions grow well above current levels, leaving the U.S. with a renewed dependence on electricity from oil and gas and a high level of SO_2 emissions.

Accelerated Coal Program — SO_2 Emission Reduction Technologies: In order to anticipate and avoid long-term environmental con-

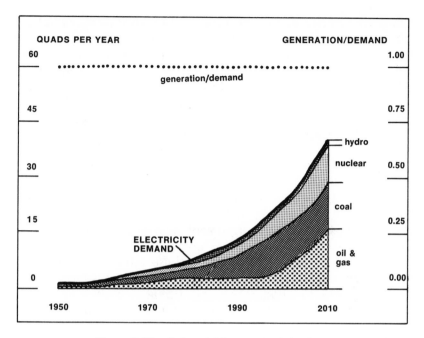

Figure 5-17. Relaxed SO$_2$ standards projection

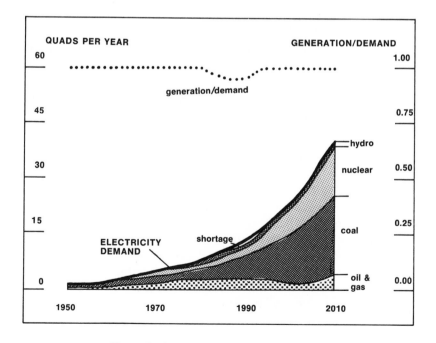

Figure 5-18. Stack gas scrubber projection

straints to coal use in utilities, Figure 5-18 tests the effects of legislation encouraging the implementation of SO_2 scrubbers on all power plants. In this simulation, all coal plants are constructed or retrofitted with stack gas devices by 1990 at a cost of 50 dollars per kilowatt (in 1970 dollars). SO_2 emissions per plant are reduced by 90 percent with the installation of each stack gas device (Battelle 1975).

Figure 5-18 shows that installing stack gas scrubbers is a more effective policy than relaxing environmental standards (Figure 5-17). The amount of oil and gas used for generating electricity remains constant near current levels to the year 2010, for most power plant capacity additions are fired by coal or nuclear power. Unlike the policy relaxing environmental standards, emission control policies allow utilities to increase coal-fired electricity without increasing total SO_2 emissions. The environment is no worse off for the growth in coal-fired capacity. Yet neither policy is free of detrimental side effects: relaxing standards allows coal-use growth at the cost of degrading the environment; installing scrubbers buys growth at the cost of the capital equipment needed. In Figure 5-18 utilities cannot acquire enough new capital to meet the higher capital needs of coal-fired plants fitted with pollution abatement equipment. From 1980 to 1990, the nation is faced with intermittent brownouts.

Combined Electricity Policies

Figures 5-13 through 5-18 suggest that the electric utility industry is close to its financial limits. An increase in electricity demand above the projected 5 percent per year growth rate or an increase in investment requirements for pollution abatement equipment could precipitate a generating capacity shortage in the 1980s. The following simulations test the combined effectiveness of financial and fuel-mix policies, carried out both with and without a moratorium on nuclear power plant construction.

Combined Financial and Fuel Mix Policies: Figure 5-19 shows the effects of combining utility financing and fuel mix policies. Rate relief and load management policies like those tested in Figure 5-15 tend to release the financial constraint on new investment. Rate relief increases the industry's ability to attract external capital, and load management alleviates some of the need for new generating capacity. Stack gas scrubbers are widely installed on existing and planned coal-fired utilities (as in Figure 5-18), releasing the environmental constraint on future investments in coal-fired generating equipment.

The combination of rate relief and load management is sufficient to avoid any shortage in electrical generating capacity over the period

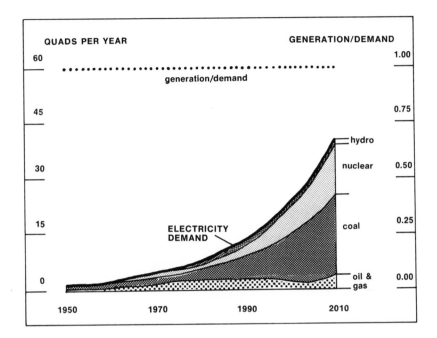

Figure 5-19. Combined financial and fuel mix projection

of the simulation, even with the increased capital requirements from the installation of pollution abatement equipment. Adding scrubbers causes electricity generated from oil and gas to drop after 1980 as both coal and nuclear power take over the burden. Nuclear power increases from 7 percent of total electricity generated in 1975 to 35 percent in the year 2010, while coal's share increases from 45 to 55 percent between 1975 and 2010.

Adding a Nuclear Moratorium: A growing body of literature has questioned the technical and ethical justification for increased dependence on nuclear power as an energy source (see, for example, Kneese 1973; Alfven 1972; Weinberg 1972). Yet many nuclear proponents claim that nuclear power is an indispensable part of "Project Independence" (Ray 1973, p. 49; ERDA 1975, p. IV-6). In order to examine the necessity for continued high rates of growth in nuclear power development, Figure 5-20 examines the behavior of the electricity sector if, in addition to the policy combination of Figure 5-19, a moratorium is placed on the construction of nuclear power plants after 1980 (those initiated before 1980 are completed). A comparison of Figures 5-19 and 5-20 shows that the curtailment of nuclear power does create some negative side effects that must be weighed against the potential benefits of abstaining from a source of

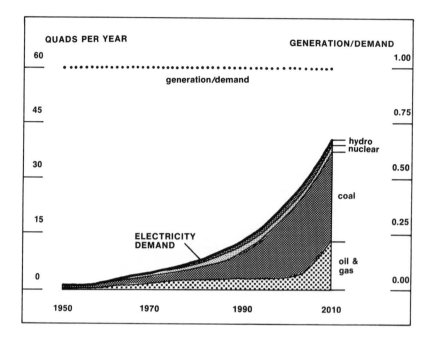

Figure 5-20. Combined policies plus nuclear moratorium

power with long-term risks: SO_2 emissions from coal (not plotted in Figure 5-20) are increased, and additional quantities of imported oil and gas must be burned after the year 2000. Yet the simulation shown in Figure 5-20 is still a significant improvement over the reference run. Because of the financial policies implemented in 1977 (rate relief and load management), the moratorium on nuclear plant construction does not create a shortage of generating capacity. The use of oil and gas as a utility fuel remains low as coal-fired plants fitted with emission control equipment take over the major burden for generating electricity. By the year 2000 about 70 percent of U.S. electricity is generated in coal-fired plants. The high rate of growth in electricity generated from coal causes SO_2 emissions to reach the maximum acceptable levels by 1990. Additional SO_2 removal technologies more efficient than the scrubbers which are implemented in the simulation (for example, solvent-refined coal or fluidized bed combustion) might be commercially available by this time to further reduce emissions per plant. With the installation of stack gas scrubbers only, oil- and gas-fired capacity begins to grow again after the year 2000.

By 2010, the increased coal use due to the nuclear moratorium requires 65 quads per year (2.7 billion tons) of coal to fuel utility boilers. Given that 1975 coal production was 15 quads per year (640 million tons), this projection implies a fourfold increase in coal production by the year 2010 from increased utility demand alone. Chapter Six (the coal sector) examines whether such high rates of growth in coal production can be achieved.

CONCLUSIONS

The reference projection of the electricity sector identifies two potential problem areas: financing and fuel mix decisions. If demand should return to historical growth rates, shortages up to 3 percent of annual demand could occur between 1975 and 1995. To avoid such shortages, utility regulatory procedures should be revamped (rate relief), and improved load management should be instituted. Figure 5-15 illustrates that if both changes are made, potential shortages of electrical generating capacity could be avoided.

The reference projection (Figure 5-13) also indicates that with current SO_2 regulations, the use of oil and gas for generating electricity will grow substantially in the future. Oil and gas demand from utilities grows to a staggering 50 quads per year in the reference projection by 2010, which is higher than the 1972 peak production of oil and gas in the United States.

The ability of nuclear power to satisfy a major portion of future electricity demand is severely limited by social, economic, and environmental considerations. Even though the electricity sector considers only economic factors (such as financing and unit cost comparisons), the reference run shows that nuclear power satisfies less than 30 percent of electricity demand by the year 2010. Even the Accelerated Nuclear projection of Figure 5-16 would allow nuclear power to satisfy only 35 percent of electricity needs by the year 2010. Under a wide range of assumptions, the COAL2 model projects that steam-electric generation from fossil fuels will be the primary source of electricity during the transition period.

The most effective measures that avoid continued growth in United States dependence on oil and gas as a fuel for fossil-fired utilities involve major policy changes to encourage the use of coal as a boiler fuel. Relaxing environmental standards only postpones the resolution of the environmental issues associated with burning coal (see Figure 5-17). A more effective policy would be a massive effort to install pollution-abatement technologies such as stack gas devices and fluidized-bed combustion as soon as feasible, while relying on the use of low-sulfur coal to reduce emissions in the interim (Figure 5-18). When SO_2 emissions technologies are combined with other policies that improve the financial health of utilities (rate relief and load management), the projected financial and fuel mix problems of the utilities are solved. The improved behavior of the electricity sector shown in Figure 5-19 resulted from the following utility policy program:

- higher allowed rate of return
- no regulatory lag in rate adjustments
- load management to achieve 60 percent capacity utilization
- stack gas scrubbers by 1990

Finally, Figure 5-20 examines the feasibility of a moratorium on the construction of nuclear power plants (tested in combination with the above financial and fuel-mix policies). Although the moratorium does impose a burden on the rest of the energy system (increasing both SO_2 emissions and the use of oil and gas), these costs may be weighed against the benefits of avoiding the long-term risks of nuclear power. A detailed evaluation of the effects of electrical energy policies on the other sectors of the energy system is included in Chapter Seven.

CHAPTER 5
EQUATIONS

```
100  *       ELECTRICITY SECTOR OF COAL2
110  NOTE
120  NOTE    ELECTRICITY SUPPLY-DEMAND BALANCE
130  NOTE
140  A       SED.K=NELD.K-HG.K
150  A       SEG.K=NEG.K+CEG.K+OGEG.K
160  A       TEG.K=SEG.K+HG.K
170  A       SEGC.K=NEGC.K+CEGC.K+OGEGC.K
180  A       SECF.K=TABHL(SECFT,SED.K/SEGC.K,0,1,.2)
190  T       SECFT=0/.2/.4/.6/.77/.85
200  A       EGDR.K=TEG.K/NELD.K
210  NOTE
220  NOTE    ELECTRIC UTILITY FINANCING
230  NOTE
240  A       EP.K=CLIP(ALELP.K,RELP.K,TIME.K,RRT)
250  C       RRT=4000
260  A       RELP.K=DLINF1(ALELP.K,RLT)
270  N       RELP=7.16E-6
280  C       RLT=2
290  A       ALELP.K=AECOST.K+AROR.K*SEC.K/SEG.K
300  A       AROR.K=CLIP(IAROR,NAROR,TIME.K,RRT)
310  C       NAROR=.08
320  C       IAROR=.10
330  A       SEC.K=NUC.K+CUC.K+OGUC.K
340  A       AECOST.K=(NELC.K*NEG.K+CEC.K*CEG.K+OGEC.K*OGEG.K)/SEG.K
350  A       SER.K=EP.K*SEG.K
360  A       FSERI.K=TABHL(FSERIT,SECF.K/SECFN.K,.7,1.1,.1)*EROIM.K
370  T       FSERIT=0/.1/.3/.55/.7
380  A       SECFN.K=CLIP(SECF2,SECF1,TIME.K,PYEAR)
390  C       SECF1=.55
400  C       SECF2=.55
410  A       EROIM.K=TABHL(EROIMT,AEROI.K,0,.10,.02)
420  T       EROIMT=.25/.3/.47/.85/1/1
430  A       AEROI.K=SMOOTH(EROI.K,EROIAT)
440  C       EROIAT=5
450  A       EROI.K=(EP.K-AECOST.K)*SEG.K/SEC.K
460  A       SEUI.K=FSERI.K*SER.K
470  NOTE
480  NOTE    HYDROPOWER GENERATION
490  NOTE
500  A       HG.K=TABHL(HGT,TIME.K,1950,2010,10)*1E15
510  T       HGT=.32/.50/.84/1.1/1.3/1.38/1.4
520  NOTE
530  NOTE    NUCLEAR ELECTRICITY GENERATION
540  NOTE
550  L       NUC.K=NUC.J+(DT)(NUICR.JK-NCDR.JK)
560  N       NUC=NUCI
570  C       NUCI=0
580  R       NCDR.KL=NUC.K/ALNU
590  C       ALNU=35
600  R       NUIR.KL=FIN.K*SEUI.K
610  R       NUICR.KL=DELAY3(NUIR.JK,NCT.K)
620  A       NCT.K=CLIP(NCT2,NCT1,TIME.K,PYEAR)
630  C       NCT1=10
640  C       NCT2=10
```

```
650 A      FIN.K=CLIP(FIN2.K,FIN1.K,TIME.K,NMYEAR)
660 A      FIN1.K=TABHL(FIN1T,FNCR.K,0,2.5,.5)
670 C      NMYEAR=4000
680 T      FIN1T=0/.25/.5/.72/.87/1
690 A      FIN2.K=TABHL(FIN2T,FNCR.K,0,2.5,.5)
700 T      FIN2T=0/0/0/0/0/0
710 A      FNCR.K=AFEC.K/NELC.K
720 A      AFEC.K=(CEC.K*CEG.K+OGEC.K*OGEG.K)/(CEG.K+OGEG.K)
730 A      NCCR.K=TABHL(NCCRT,TIME.K,1950,2010,10)*1E6
740 T      NCCRT=1E-6/1E-6/.062/.056/.051/.051/.051
750 A      NFC.K=TABHL(NFCT,TIME.K,1950,2010,10)*1E-6
760 T      NFCT=.44/.44/.44/.49/.78/1.23/1.95
770 A      NELC.K=UCCAF/(NCCR.K*CFCC.K)+NFC.K
780 C      UCCAF=.14
790 A      CFCC.K=CLIP(CFCC2,CFCC1,TIME.K,PYEAR)
800 C      CFCC1=.7
810 C      CFCC2=.7
820 A      NEGC.K=NCCR.K*NUC.K
830 A      NEG.K=SECF.K*NEGC.K
840 NOTE
850 NOTE   ELECTRICITY GENERATION FROM COAL
860 NOTE
870 L      CUC.K=CUC.J+(DT)(CUICR.JK-CUDR.JK)
880 N      CUC=FFUCI*FFUCCI
890 C      FFUCI=23.8E9
900 C      FFUCCI=.66
910 R      CUDR.KL=CUC.K/ALFFU
920 C      ALFFU=35
930 R      CUIR.KL=FICU.K*IFFU.K
940 R      CUICR.KL=DELAY3(CUIR.JK,FFUCT)
950 C      FFUCT=5
960 A      IFFU.K=(1-FIN.K)*SEUI.K
970 A      FICU.K=CLIP(FICU2.K,FICU1.K,TIME.K,PYEAR)
980 A      FICU1.K=TABHL(FICU1T,FCR.K,0,2,.5)*IMES.K
990 T      FICU1T=0/.15/.66/.88/1
1000 A     FICU2.K=TABHL(FICU2T,FCR.K,0,2,.5)*IMES.K
1010 T     FICU2T=0/.15/.66/.88/1
1020 A     FCR.K=OGUFC.K/CUFC.K
1030 A     CUFC.K=CPRICE.K/FFECE.K
1040 A     FFECE.K=TABHL(FFECET,TIME.K,1950,2010,10)
1050 T     FFECET=.24/.31/.32/.33/.34/.35/.36
1060 A     FFCCR.K=TABHL(FFCCRT,TIME.K,1950,2010,10)*1E6
1070 T     FFCCRT=.064/.082/.103/.084/.084/.084/.084
1080 A     CEC.K=UCCAF/(CFCC.K*FFCCR.K)+CUFC.K
1090 A     SO2E.K=CEG.K*SO2EF.K
1100 A     SO2EF.K=TABHL(SO2EFT,TIME.K,1950,2010,10)*1E-9
1110 T     SO2EFT=6.1/6.1/5/2/2/2/2
1120 A     SO2ESD.K=CLIP(SO2ESD2,SO2ESD1,TIME.K,PYEAR)
1130 C     SO2ESD1=6.9E6
1140 C     SO2ESD2=6.9E6
1150 A     IMES.K=TABHL(IMEST,SO2ESD.K/SO2E.K,0,1.2,.2)
1160 T     IMEST=0/.04/.14/.38/.92/1/1
1170 A     CEGC.K=CUC.K*FFCCR.K
1180 A     CEG.K=SECF.K*CEGC.K*CPDR.K
1190 A     CDU.K=SECF.K*CEGC.K/FFECE.K
1200 NOTE
1210 NOTE  ELECTRICITY GENERATION FROM OIL AND GAS
1220 NOTE
1230 L     OGUC.K=OGUC.J+(DT)(OGUICR.JK-OGUDR.JK)
1240 N     OGUC=FFUCI*(1-FFUCCI)
1250 R     OGUDR.KL=OGUC.K/ALFFU
1260 R     OGUIR.KL=IFFU.K*(1-FICU.K)
1270 R     OGUICR.KL=DELAY3(OGUIR.JK,FFUCT)
1280 A     OGUFC.K=AOGP.K*OGUFCF/FFECE.K
```

```
1290 C        OGUFCF=.54
1300 A        OGEC.K=UCCAF/(CFCC.K*FFCCR.K)+OGUFC.K
1310 A        OGEGC.K=FFCCR.K*OGUC.K
1320 A        OGEG.K=SECF.K*OGEGC.K*OGCDR.K
1330 A        OGDU.K=SECF.K*OGEGC.K/FFECE.K
1340 NOTE
1350 NOTE     SUPPLEMENTARY EQUATIONS
1360 NOTE
1370 A        COGEG.K=OGEG.K+CEG.K
1380 A        COGNEG.K=COGEG.K+NEG.K
1390 A        FGOG.K=OGEG.K/TEG.K
1400 A        FGCOG.K=COGEG.K/TEG.K
1410 A        FGCOGN.K=COGNEG.K/TEG.K
1420 A        OGPDU.K=AOGP.K*OGUFCF
1430 A        FSECN.K=NEGC.K/SEGC.K
1440 A        FFFCC.K=CEGC.K/(CEGC.K+OGEGC.K)
1450 A        FINC.K=FIN.K+(1-FIN.K)*FICU.K
1460 NOTE
1470 NOTE     EXOGENOUS INPUTS
1480 NOTE
1490 L        NELD.K=NELD.J+(DT)(ELDCR.JK)
1500 N        NELD=NELDI
1510 C        NELDI=1.15E15
1520 R        ELDCR.KL=NELD.K*ELDGR.K
1530 A        ELDGR.K=CLIP(ELGR1.K,LTELGR.K,TIME.K,RYEAR)
1540 C        RYEAR=1973
1550 A        LTELGR.K=TABLE(LTELGRT,TIME.K,1950,2010,10)*1E-2
1560 T        LTELGRT=7.5/7.5/7.5/5.4/5.4/5.4/5.4
1570 A        ELGR1.K=CLIP(LTELGR.K,RELGR.K,TIME.K,RCYEAR)
1580 C        RCYEAR=1976
1590 A        RELGR.K=TABHL(RELGRT,TIME.K,1973,1975,1)*1E-2
1600 T        RELGRT=-2/-2/2
1610 A        AOGP.K=TABLE(AOGPT,TIME.K,1950,2010,10)*1E-6
1620 T        AOGPT=.65/.65/.65/1.5/2/2.1/2.1
1630 A        CPRICE.K=TABLE(CPRICET,TIME.K,1950,2010,10)*1E-6
1640 T        CPRICET=.35/.35/.35/.5/.6/.6/.6
1650 A        OGCDR.K=TABLE(OGCDRT,TIME.K,1950,2010,10)
1660 T        OGCDRT=1/1/1/1/1/1/1
1670 A        CPDR.K=TABLE(CPDRT,TIME.K,1950,2010,10)
1680 T        CPDRT=1/1/1/1/1/1/1
1690 NOTE
1700 NOTE     CONTROL CARDS
1710 NOTE
1720 N        TIME=1950
1730 C        PYEAR=1977
1740 SPEC     DT=.2/LENGTH=0/PLTPER=2/PRTPER=0
1750 PLOT     TEG=G,NELD=D,COGNEG=N,COGEG=C,OGEG=O(0,60E15)/EGDR=*(.8,1)
1760 PRINT    FSECN,FFFCC,EP
1770 RUN
1780 NOTE
1790 NOTE     PARAMETER CHANGES FOR THE ELECTRICITY SECTOR RUNS
1800 NOTE
1810 NOTE     HISTORICAL RUN
1820 NOTE
1830 C        LENGTH=1975
1840 C        PLTPER=1
1850 PLOT     EROI=R(0,.12)/EP=$(0,10E-6)
1860 PLOT     FIN=N,FINC=C(0,1)
1870 PLOT     FGOG=O,FGCOGN=N,FGCOG=C(0,1)
1880 PLOT     TEG=G,NELD=D,COGNEG=N,COGEG=C,OGEG=O(0,1E16)
1890 RUN      HISTORICAL RUN
1900 NOTE
1910 NOTE     REFERENCE RUN
1920 NOTE
```

```
1930 CP    LENGTH=2010
1940 PLOT   EROI=R(0,.4)/EP=$(0,10E-6)
1950 PLOT   TEG=G,NELD=D,COGNEG=N,COGEG=C,OGEG=O(0,60E15)/EGDR=*(.8,1)
1960 RUN    REFERENCE RUN
1970 NOTE
1980 NOTE   POLICY RUNS
1990 NOTE
2000 NOTE   FINANCIAL POLICIES
2010 NOTE
2020 T      LTELGRT=7.5/7.5/7.5/7.5/7.5/6/4.5
2030 RUN    INTENSIVE ELECTRIFICATION PROJECTION
2040 T      LTELGRT=7.5/7.5/7.5/7.5/7.5/6/4.5
2050 C      RRT=1977
2060 T      SECFT=0/.2/.4/.6/.8/.85
2070 C      SECF2=.60
2080 C      CFCC2=.75
2090 RUN    RATE RELIEF & LOAD MANAGEMENT
2100 NOTE
2110 NOTE   FUEL MIX POLICIES
2120 NOTE
2130 PLOT   TEG=G,NELD=D,COGNEG=N,COGEG=C,OGEG=O(0,60E15)/EGDR=*(.8,1)
2140 C      NCT2=6
2150 T      NFCT=.44/.44/.44/.49/.64/.82/1.07
2160 RUN    ACCELERATED NUCLEAR PROGRAM
2170 C      SO2ESD2=10.4E6
2180 RUN    RELAXATION OF SO2 STANDARDS
2190 T      SO2EFT=6.1/6.1/5/2/.6/.6/.6
2200 T      FFCCRT=.064/.082/.103/.084/.073/.073/.073
2210 RUN    SO2 EMISSIONS TECHNOLOGIES
2220 C      RRT=1977
2230 TP     SECFT=0/.2/.4/.6/.8/.85
2240 CP     SECF2=.60
2250 CP     CFCC2=.75
2260 TP     SO2EFT=6.1/6.1/5/2/.6/.6/.6
2270 TP     FFCCRT=.064/.082/.103/.084/.073/.073/.073
2280 RUN    COMBINED FINANCIAL AND FUEL MIX POLICIES
2290 C      NMYEAR=1980
2300 T      NFCT=.44/.44/.44/.49/.7/.75/.8
2310 RUN    NUCLEAR MORATORIUM
```

The Coal Sector

PURPOSE OF THE COAL SECTOR

The coal sector of the COAL2 model is designed to examine the long-term dynamics of coal supply in the United States. The demand, oil and gas, and electricity sectors of COAL2 determine the demand for coal from direct-use consumption, synthetic fuel facilities, and coal-fired utilities respectively. The coal sector must attract sufficient capital and labor inputs to meet changing patterns of demand.

In the coal sector equations, coal costs are linked to technological changes, labor and capital costs, health and safety legislation, and resource depletion effects. In addition, coal prices can greatly exceed costs when coal demand expands faster than existing production capacity. Because the capital and labor requirements, resource base, and environmental considerations differ greatly between strip and underground mines, the coal sector includes both a strip and an underground production subsector.

As noted by the FEA Project Independence Blueprint (FEA-CTF 1974, p. 1), the ability of the United States coal industry to meet an accelerated demand for coal is critical to any plan for United States independence from foreign oil. The basic problem facing the United States coal industry is unique. For the past 60 years United States coal production has remained near 600 million tons per year. Yet, in order to balance future domestic demand with supplies, coal production may have to increase by as much as 10 percent per year (the FEA "Accelerated Development" scenario; FEA-CTF 1974, p. 16) during the next three decades. The coal sector of COAL2 can be used to analyze the ability of the coal industry to move from stagnation to high growth rates, and to design policies that encourage coal industry growth.

When the coal sector of the model is run, two difficulties arise which have the potential for disrupting the industry's response to increasing demand:

1. A *startup problem*, in which the industry is unable to increase its output fast enough to keep up with accelerating demand.
2. A *depletion problem*, in which the depletion of surface resources creates an accelerated demand for underground coal that exceeds the expansion capability of the underground coal industry.

Government intervention can either help or hinder the normal market mechanisms that attempt to balance coal supply and demand. The coal sector structure also includes the capability to test the following legislative policy options:

- loan guarantees
- price supports
- 1969 Coal Mine Health and Safety Act
- ban on surface coal mining
- surface mining restrictions (slope restrictions, surface coal tax, reclamation legislation)

The COAL2 model analysis concludes that both financial incentives (price guarantees) and restrictions on surface mining are required to meet projections of accelerated coal demand.

HISTORICAL BEHAVIOR

Figure 6-1 illustrates the historical behavior of United States coal consumption (by usage category) from 1950 to 1974. The total amount of coal consumed in the United States has varied between 400 and 600 million tons per year (10-15 quads per year). During the historical period, direct coal-use demand (coking coal, general industry, and retail coal) dropped substantially, due to a general shift to petroleum and natural gas in end-use markets (see Chapter Three). The use of coal as a boiler fuel in steam-electric generating plants grew from less than 20 percent to about 65 percent of total coal consumption during the 23-year period (Figure 6-2). Exports remained constant near 1.5 quads per year (60 million tons), equal to 10 percent of total coal use.

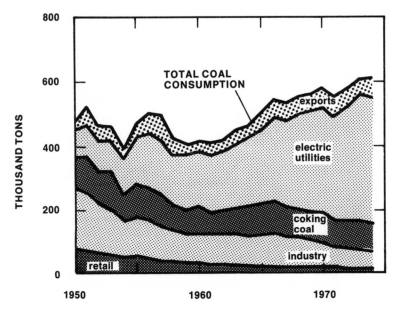

Source: Coal Data 1974, p. 82.

Figure 6-1.   Coal demand by end use

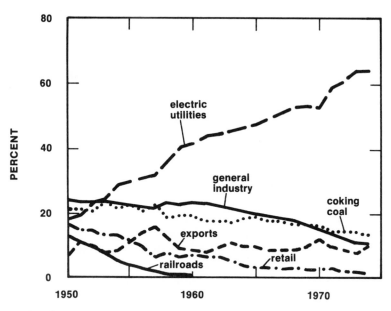

Source: Coal Data 1974. Used by permission.

Figure 6-2.   Coal end-use categories as a percent of consumption

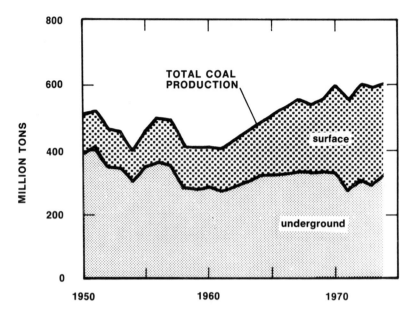

Source: Data from Conoco 1973, p. 16; FEA 1976, p. 172.

Figure 6-3.   Coal production: underground vs. surface

Figure 6-3 shows the production of coal in the United States by mining method (surface or underground). Surface-mined production has grown from 24 percent of total production in 1950 to 54 percent in 1974 (*Coal Facts* 1974, p. 53; FEA 1976, p. 172).

Figure 6-4 demonstrates the behavior of the coal labor force over the period of 1950-1974. Employment in underground mines dropped sharply from 375,000 men in 1950 to 111,000 men in 1973, while surface mine employment has remained relatively constant, near 30,000 men, during the period. The major reason for the declining underground labor force was a general industry shift towards more capital-intensive mining methods (noted as continuous mining and longwall mining in Figure 6-5).

Underground mine safety and wages will play a critical role in determining whether the coal industry can attract sufficient labor in the future. Historically, the United States underground coal industry has had the worst safety record of any major industry, averaging approximately one fatality and 50 nonfatal injuries per million man-hours (four times the injury rate of comparable industries; *Coal Facts* 1974, p. 89). Yet, as shown in Figure 6-6, since enactment of the 1969 Mine Health and Safety Act, underground coal mine safety has improved by a factor of about two. Coal wages have historically

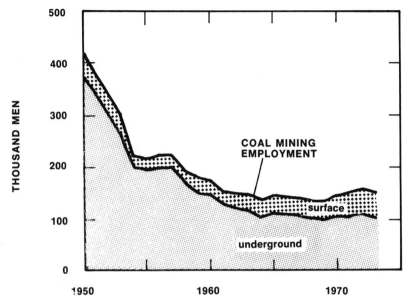

Source: Coal Data 1974, p. 33. Used by permission.

Figure 6-4. Coal employment: underground vs. surface

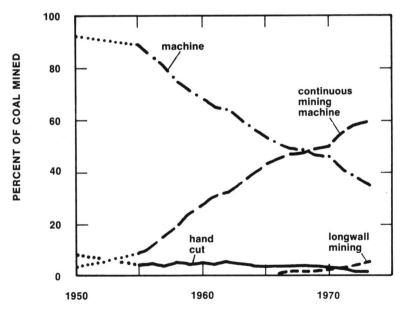

Source: Coal Data 1974, p. 53. Used by permission.

Figure 6-5. Underground production by method

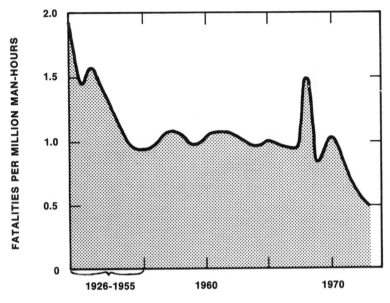

Source: *Coal Facts* 1974, p. 90.

Figure 6-6.   Underground safety record

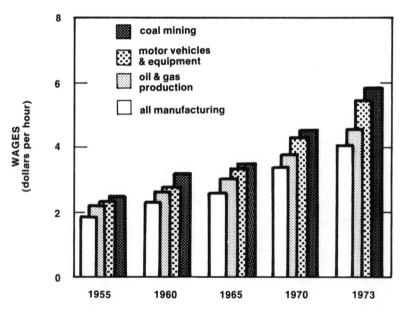

Source: *Coal Data* 1974, p. 45. Used by permission.

Figure 6-7.   Coal mining wages compared to other industries (bituminous mining only)

been higher than the norm (Figure 6-7), partially offsetting the exceptional risk of underground mining.

Figure 6-8 indicates that the total capitalization of the coal industry (including transportation) grew substantially from around 3.9 billion dollars in 1950 (in 1970 dollars) to 5.7 billion dollars in 1970. Yet most of this capital expansion went toward the replacement of men with machines (Figures 6-4 and 6-8): the total production capacity of the industry has remained almost constant. Figure 6-9 shows the historical behavior of production capacity, based on 250 operating days maximum per year (Risser 1973, p. 12). Coal producers can vary the average number of days worked in any given year to adjust coal production to meet demand. The increase in average number of days worked from 1960 to 1973 indicates that production capacity has not kept pace with recent increases in demand.

The average delivered price of coal to steam-electric plants (Figure 6-10) has remained essentially constant (in constant dollars) during the historical period. Apparently coal resource depletion has had little (if any) effect on production costs. Coal prices began to rise after 1970 due primarily to cost increases from the 1969 Coal Mine Health and Safety Act (CRS 1973, p. 23; NPC 1973, p. 44).

If the coal industry is to meet accelerated demands for coal, a number of its historical trends must be sharply reversed. The most obvious change must occur in underground coal mine employment. There, the industry must reverse its history of declining employment and attract large numbers of new miners in order to expand. Coal capacity, currently increasing at about 2 percent per year, must accelerate to 10 percent per year over the long term to meet the FEA "Accelerated Development" scenario (FEA-CTF 1974, p. 16).

BASIC CONCEPTS

Composition of the Coal Industry

The United States coal industry is divided into two coal-producing subsectors in the COAL2 model: surface-mined coal and underground coal. Each subsector has a separate capital stock, resource base, production process, and production cost. However, when surface-mined and underground coal are "delivered" to the coal-consuming sectors of the COAL2 model (industrial coal users, synthetic conversion facilities, or coal-fired utilities), the delivered product is homogeneous. A Btu of surface or underground coal is delivered at an average coal price, which includes an average delivery cost.

In 1973, 14 percent of total coal production came from captive mines owned by steel companies, public utilities, or other industries

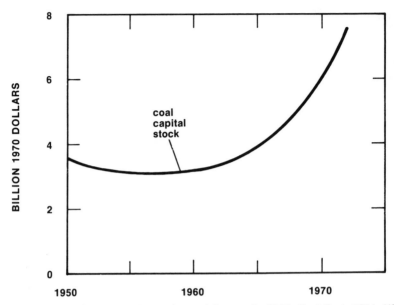

Source: Data from Creamer, Dobrovolsky, and Borenstein 1960; *Coal Facts* 1974; NPC 1973; *Coal Data* 1973.

Figure 6-8.   Capitalization of the coal industry

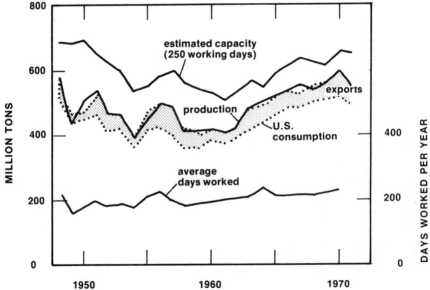

Source: Risser 1973, p. 12. Used by permission.

Figure 6-9.   Coal production, capacity, and days worked

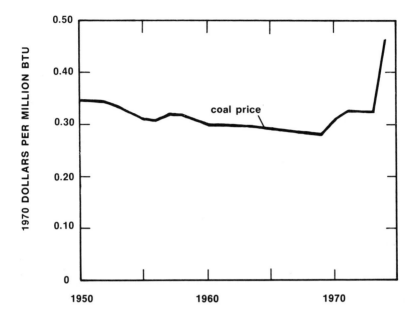

Source: Data from EEI 1974a, p. 50 (and earlier years).

Figure 6-10.   Cost of coal burned for electricity

(Keystone 1974, p. 685). The fifty biggest mines produced 25 percent of total coal production in 1973; of these mines 7 were captive (Keystone 1974, p. 447). In 1974 coal production by companies controlled by petroleum and conglomerate firms account- ed for about 40 percent of total output. Although there are very few "pure" coal companies, the companies comprising the industry are in fact separate financial entities and are treated as such in the coal sector of COAL2. For example, any large transfer of funds from a petroleum company to an associated coal company is considered financing external to the coal company, both in the model and in the actual coal company financial statements.

Coal Production Functions

A production function is a mathematical expression relating one or more inputs (factors of production) to an output (product). The coal sector employs two production functions: one for surface-mined coal and one for underground coal. Both can be described as modified Cobb-Douglas production functions (see Ervik 1974). An analytical expression of the production functions used in COAL2 would take the following form:

$$X = g * K^a L^b R^c \tag{6-1}$$

where X = coal output
          K = capital
          L = labor
          R = resource stock

A strictly defined Cobb-Douglas production function would define a,b,c (the input elasticities) and g (a productivity measure) as constants. The form used in COAL2 employs variable, nonlinear elasticities and therefore is a modified Cobb-Douglas production function.

When a factor of production (K,L,R) is dropped from the production function, the implicit assumption is that coal output would not be affected by a change in the deleted factor—the input elasticity is zero. Most studies leave resources out of their production functions (NPC 1973, p. 17; FEA-CTF 1974, p. 43). COAL2 includes the effects of resource depletion on both surface and underground production. Because labor availability is not expected to constrain surface coal production even under an accelerated growth scenario (MITRE 1975, pp. 5-10), labor is omitted from the surface-mining production function in COAL2.

Historically, technology has changed the capital-labor ratio in the coal production function (Ervik 1975, p. 16). The effects of technology manifest themselves in statistics of *productivity* (output per man-day). Productivity has been increasing in both the surface and underground industries as they have become more capital-intensive (NPC 1973, pp. 36-37). Because labor is not included in the surface-mining production function, the effect of technology on the surface capital-labor mix is also absent. Since technology has been a major factor in explaining the historical behavior of underground coal production (the shift to capital-intensive mining methods), the underground coal production function does include a technological trend.

Policy changes such as the 1969 Coal Mine Health and Safety Act can also affect coal production. The effect of the 1969 Act on the underground coal production function is modeled in COAL2 as a direct decrease in underground productivity (the coefficient g in Equation 6.1). Up to 25 percent more input of capital (in the form of safety equipment) and labor (in the form of inspectors and safety equipment operators) is required per unit of output due to new safety regulations (CRS 1973, p. 23). Environmental restrictions on surface mining can also affect coal production, increasing unit costs.

The effects of both the 1969 Health and Safety Act and federal surface-mining legislation are tested in the simulation runs.

The underground and surface production functions determine the *production capacity* of existing coal mines in the COAL2 model. Production capacity establishes an upper limit to coal production—producers normally operate their mines well below the maximum capacity, adjusting capacity utilization (days worked per year) to meet demand. If the United States chooses to accelerate the use of coal during the transition period, sufficient amounts of capital, labor, and resources must be available to expand production capacity rapidly enough to keep ahead of demand. The structure of the coal sector therefore focuses on the basic mechanisms that control the flows of capital, labor, and resources in the United States coal industry.

Coal Capital Availability

A number of studies have suggested that, because total capital requirements for the coal industry are small when compared to other energy industries, there will be no coal capital shortage (Hass, Mitchell, and Stone 1974, p. 97; MITRE 1975, p. 73). But a Dartmouth substudy by Ervik (1975) argues that the *timing* of coal capital availability may present a real problem. To avoid future coal shortages in mine capacity, coal investment must anticipate (rather than react to) large increases in demand. Anticipation is a must because the opening of new mines normally requires lead times of almost five years.

The financial structure of the coal sector appears in Figure 6-11. The need for new investment is measured by the ratio of coal demand to capacity. As coal demand accelerates, profits and return on investment increase, thus creating the incentive for increased investment in the industry. Yet coal investments (the fraction of coal revenues invested multiplied by revenues in Figure 6-11) do not respond immediately to an increase in the rate of return. Investors are reluctant to change their past investment patterns due to the uncertainties of the future, such as new air pollution legislation or changes in the world oil price (FEA-CTF 1974, p. 59). Even after substantial increases in investment, a 3- to 5-year lead time is required to open a new mine (NAE 1974, p. 92).

Figure 6-11 shows that an increase in coal mine capacity can lead to an increase in coal production (if warranted by increased demand). The increased production generates new revenues, thereby forming more capital in the course of time. As the investment

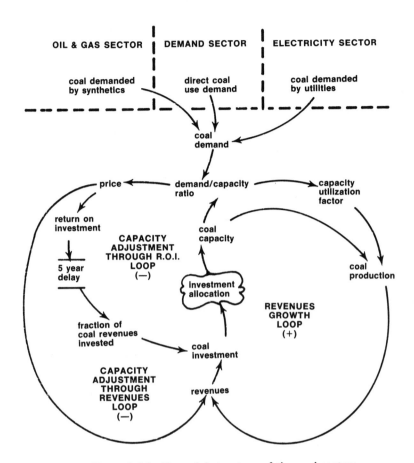

Figure 6-11. Financial structure of the coal sector

matures into new production facilities, a positive feedback loop (revenues growth loop in Figure 6-11) is closed. The loop embodies the industry growth potential from capital accumulation. Two negative feedback loops (capacity adjustment through ROI and capacity adjustment through revenues) tend to moderate industry growth through the operation of the price mechanism. If capacity exceeds demand, coal prices decrease, which reduces investment by decreasing both revenues and return on investment. The reduced investment eventually adjusts coal production capacity to demand, thereby completing the negative feedback loops.

The delay between the first indication of a real need for new coal capacity and installation of the needed capacity could afflict the industry with a period of inadequate capacity if coal demand accelerates quickly. Policies such as price guarantees (which increase profits) or subsidies and guaranteed loans (which increase the industry's external financing ability) might alleviate the financial shortage if implemented in anticipation of increased coal demand. These policies are tested in the simulation runs presented later in this chapter.

Underground Labor Availability

Almost every major study of coal supply has concluded that manpower availability could be a restraining factor in the development of the underground coal mining industry (MITRE 1975, p. 5-5; NPC 1973, p. 48; FEA-CTF 1974, p. 47). A Penn State study foresees no manpower shortage, but only because its estimates of future coal demand and productivity lead to manpower requirements substantially below those of the MITRE and FEA studies (Penn State 1973, pp. 1-2, 1-12; FEA-CTF 1974, p. 34; MITRE 1975, pp. 5-10, 5-11).

Although projecting manpower *requirements* is a relatively straightforward process, all of the manpower studies agree on the extreme difficulties of projecting future *availability* of labor for underground mining. The Penn State study observes that a statistical analysis of labor force availability based on historical data would be meaningless if increasing labor force requirements are projected, for historically the labor force has declined (Penn State 1973, p. 1-14). Basing long-term projections on the 1968-1973 increase in underground labor from 103,000 to 111,000 men, the MITRE study gives perhaps the best analysis of manpower availability (MITRE 1975, p. 5-A-9). Both the Penn State and MITRE studies agree that the key policy levers that control labor availability are wages and working conditions (safety), yet neither study attempts to quantify the influence of these factors (Penn State 1973, p. 1-14; MITRE 1975, p. 5-8).

Although the exact dependence of hiring on safety and wages was extremely difficult to quantify, the coal sector explicitly includes a causal structure controlling labor availability (shown in Figure 6-12). The COAL2 model employs the concept of a *hiring adjustment time* in calculating the hiring rate, or actual number of men attracted and added to the underground work force in any given year. Underground labor requirements (determined by the number of men needed to operate the mining equipment in place) are compared to the current underground labor supply to determine the number of miners the industry needs to hire. Historically, the labor needed has been less than the labor available, and the labor force has consequently been declining (see Figure 6-4). Therefore, the hiring adjustment time was equal to the short time required to lay off a worker.

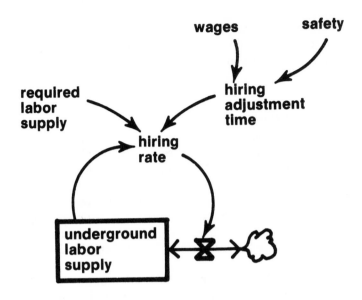

Figure 6-12. Underground labor availability structure

Under conditions of accelerated demand, however, the required labor supply for underground mines can be expected to grow considerably. When miners must be attracted to the industry, the hiring adjustment time could become substantial. Increased wages or improved safety conditions tend to decrease the hiring adjustment time, thereby reducing shortages in manpower in periods of rapid growth in coal demand. A policy change such as the 1969 Coal Mine Health and Safety Act has the beneficial effect of increasing long run underground labor availability in COAL2, even though productivity is decreased over the short run. The net effect of the Health and Safety Act on coal production is examined later in this chapter.

	Trillion Tons	*Quads*
Mapped explored: 0-3,000' overburden	1.56	32,600
Probable additional resource in unmapped and unexplored areas: 0-3,000' depth	1.31	26,500
3,000-6,000' depth	0.34	6,900
Total	3.21	66,000

Source: USGS, 1973, p. 137.

Figure 6-13. Total United States coal resources

Coal Resources

Coal is by far the largest United States energy resource, comprising almost 75 percent of total remaining ultimately recoverable United States energy resources (*Coal Facts* 1974, p. 7). The total coal resource base is estimated at 65,000 quads (3.2 trillion tons) by the United States Geological Survey. This figure can be broken down as shown in Figure 6-13.

Figure 6-14 shows the distribution of the 1.56 trillion tons (32,600 quads) of mapped and explored coal by seam thickness and by depth. Of this total, only 2,600 quads—less than 10 percent—are classified as mineable by surface methods (USBM 1974, p. 3).

The depletable stocks of surface and underground resources included in the coal sector are best described as *recoverable resources:* those resources recoverable with existing mining technologies, under *any* foreseeable economic conditions. Existing mining methods normally permit only 50 percent of underground coal

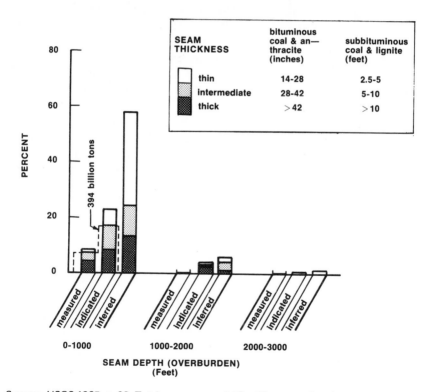

Source: USGS 1967, p. 38. Total resources are 1.56 trillion tons of coal.

Figure 6-14. Estimated coal resources, by depth and seam thickness

resources and 90 percent of surface resources to be recovered. About half the coal is left in underground mines to support the roof of the mine; only 10 percent of the coal is wasted in surface mining. (See NPC 1973, pp. 22, 23). Using the conservative mapped and explored estimate of 32,600 quads for total coal in place, the total recoverable resource base used in the coal sector is set at 17,400 quads (833 billion tons). This figure has been obtained as shown in Figure 6-15.

According to most coal supply studies, coal resources are so extensive that their depletion will have no significant effect on production costs in the foreseeable future (NPC 1973, p. 17; FEA-CTF 1974, p. 43). But accelerated rates of growth in coal production could increase coal production costs over the long term as the industry is forced to mine thinner seams located at greater depths. The National Petroleum Council estimates that only 2,200 quads (105 billion tons) of underground and 840 quads (45 billion tons) of surface coal resources are recoverable at historical production costs. Given a hypothetical annual growth rate in total coal production of 10 percent per year (the FEA "Accelerated Development" scenario), these resources would last for 29 or 21 years respectively if each alone were to satisfy all future coal demand. Because these hypothetical lifetimes fall within the time horizon of the COAL2 model (35 years), the coal sector of COAL2 links the depletion of both surface and underground coal resources to their production.

Through an econometric analysis of cross-sectional deep mine data, Martin Zimmerman of the MIT Energy Laboratory has established a relationship between underground coal costs and seam thickness (Zimmerman 1975, p. 155). When Zimmerman's relationship is combined with estimates of the fraction of the underground

Resources in Place	X	Recovery Factor	=	Recoverable Resources
Surface[a]				
137 billion tons	X	0.9[c]	=	123 billion tons
(2,600 quads)				(2,300 quads)
Underground[b]				
1,420 billion tons	X	0.5[c]	=	710 billion tons
(30,000 quads)				(15,000 quads)
Total Recoverable Resources:				833 billion tons
				(17,400 quads)

Sources: [a]USBM 1974, p. 3. [b]USGS 1973, p. 137. [c]NPC 1973, pp. 22-23.

Figure 6-15. Recoverable coal resources, as of 1974

coal resource base at each seam thickness (derived from Figures 6-14 and 6-15), the nonlinear relationship shown in Figure 6-16 results. Figure 6-16 implies that if underground coal resources were depleted to the point where "intermediate" rather than "thick" seams were being mined, twice as much capital and labor would be required to mine the same output, thereby doubling the costs of production.

A similar relationship for surface coal is shown in Figure 6-17. Surface coal resources are categorized by the stripping ratio, or overburden-to-seam-thickness ratio (feet/feet). This factor is linked to production costs using model mine cost data from the U.S. Bureau of Mines (USBM 1972, p. 3). Figure 6-17 implies that the first 25 percent of the surface coal resource base (30 billion tons of reserves) can be mined at historical costs. As the 45 billion tons of reserves are mined, productivity begins to fall, causing an increase in surface-mined coal costs.

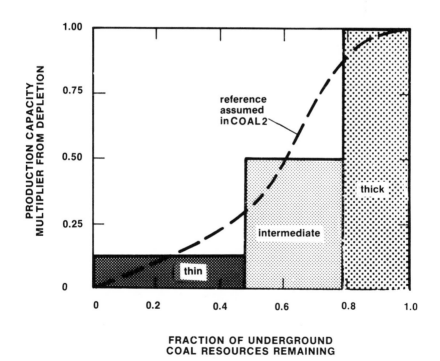

**FRACTION OF UNDERGROUND
COAL RESOURCES REMAINING**

Source: Derived from Figures 6-14, 6-15, and Zimmerman 1975, p. 156.

Figure 6-16.   Relation of underground coal depletion to production

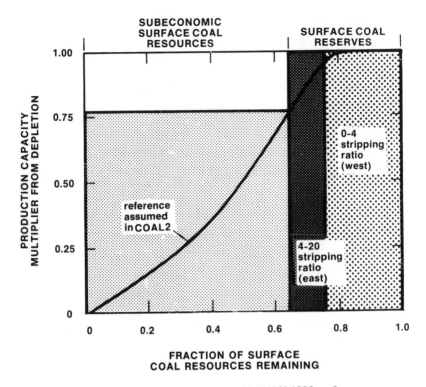

Source: Data from CEQ 1973, p. 4; USBM 1971, p. 23; USBM 1972, p. 6.

Figure 6-17.   Relation of surface coal depletion to production

COAL SECTOR CAUSAL STRUCTURE

Figure 6-18 illustrates the causal loop structure controlling surface and underground coal production in COAL2. Coal investment is divided between surface and underground production facilities by comparing their unit costs of production. Investment accumulates in a stock of production capital (surface capital and underground capital in Figure 6-18). These capital stocks are the main input to the surface and underground coal production functions modeled in COAL2.

In addition to capital, the underground coal production function also includes labor as a factor of production. The required underground labor supply is jointly established by the amount of underground capital and the required labor/capital ratio. The latter is a technology-dependent variable determined by the existing mix of mining methods in use. The lag in the adjustment of the labor supply

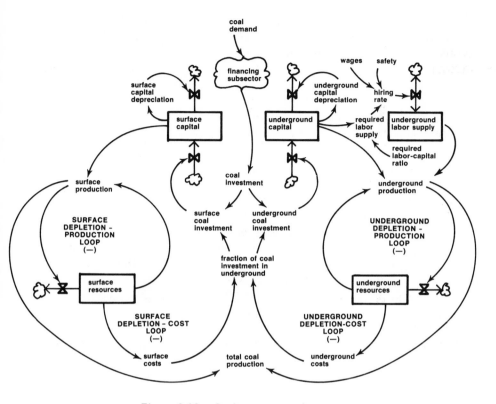

Figure 6-18.   Coal sector causal structure

to manpower requirements depends on existing wages and safety conditions.

Figure 6-18 contains four major negative feedback loops, each associated with the effects of coal resource depletion. The surface and underground depletion-production loops tend to reduce production per unit of capital or labor as resources are depleted, reflecting a decrease in productivity as thinner and deeper seams are mined. The surface and underground depletion-cost loops tend to shift investment away from surface or underground coal mining as their costs escalate, in an attempt to minimize production costs. All four loops act to ensure that the model cannot extract more coal than permitted by the physical limits of the finite stocks of surface and underground resources.

The causal structure of the coal sector includes a number of delays and negative feedback loops that might cause coal supply to lag behind demand if coal demand grows rapidly in the future. The delays in obtaining sufficient financing could cause a short-term startup problem as the coal sector attempts to reach high growth rates. Delays in capacity acquisition (mine construction delays, labor hiring delays) could cause supply to lag behind demand in a period of high growth in demand. The negative feedback loops linking coal resources to production could create a long-term coal shortage if surface or underground coal reserves are mined to depletion. The resulting growth capability of the U.S. coal industry is examined in the coal sector simulation runs described in the following section.

COAL SECTOR SIMULATION RUNS

Coal's future in the United States energy system is best described in 1977 as uncertain. The most pessimistic forecasts of coal consumption project only a 2-3 percent per year growth rate in coal use, a consequence of continued environmental restrictions on coal burning and the lack of a synthetic fuels industry (ERDA 1975, p. V-4). Yet if a federal program were undertaken to accelerate the use of coal in both utilities and synthetic conversion plants, coal consumption could grow by as much as 7.5 percent per year over the long term (FEA 1976, p. 207).

As will be seen in Chapter 7, both extremes in coal demand are generated (under different policy assumptions) in the COAL2 model. The simulations discussed in this chapter are run with exogenous projections of demand in order to examine the ability of the coal supply industry to grow. In each simulation, note first the coal production-demand ratio. If this ratio falls below its normal value of 1.0, it indicates a coal shortage. Additional government or industry policies are then necessary to accelerate the production of coal to meet demand.

These simulations indicate that the coal industry is capable of sustaining low growth rates in coal production with no government intervention. Under conditions of accelerated coal demand, however, additional policy incentives may be necessary to avoid future coal shortages. Simulations of the entire COAL2 model in Chapter Seven will examine the policies that lead to low or accelerated demands for United States coal.

Historical Behavior

To examine the model's ability to reproduce history, Figure 6-19 shows the coal sector run from 1950 to 1975. Coal demand follows a U-shaped path from 1950 to 1970, falling from 14.6 to 11.4 quads per year from 1950 to 1960, and rising back to 15 quads per year in 1970 (compare with Figure 6-3; the 1970 production of 15 quads

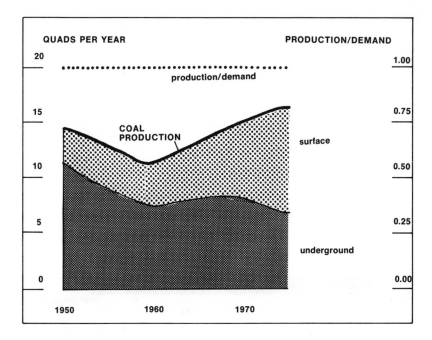

Figure 6-19. Historical behavior of the coal sector

per year is approximately equal to 600 million tons per year). Because of the lower costs of surface coal production, underground production declines by about 30 percent, while surface production increases by a factor of two. As a result, surface-mined coal increased from 25 percent of total coal production in 1950 to over 50 percent in 1975.

The effects of the 1969 Coal Mine Health and Safety Act are also evident in Figure 6-19. Underground production declines after 1970 due to a loss in production capacity from the new operating procedures outlined in the Act. Underground coal prices (not shown) increase by 10 cents per million Btu (equal to 2.40 dollars per ton) after 1970 in the model, the same as the actual impact reported from real underground mines (Conoco 1973, p. 19; CRS 1973, p. 23). The underground cost increase impairs underground coal's competitive position with respect to surface-mined coal, which further reduces underground production. A later simulation will examine the long-term effects of the 1969 Health and Safety Act.

Reference Projection

With no major changes in energy policy, coal consumption is likely to increase slowly over the long term. The Federal Energy Adminis-

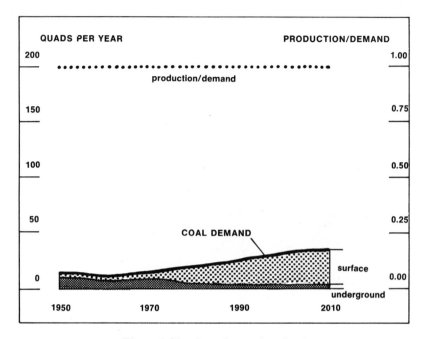

Figure 6-20.   Low demand projection

tration's *1976 National Energy Outlook* indicates that coal consumption will be 24 quads per year (1 billion tons) in 1985, a 5 percent per year growth rate for the 1974-1985 period (FEA 1976, p. 175). The 1975 ERDA plan suggests that coal use could grow even more slowly after 1985. In the ERDA Base Case, coal consumption grows to 34 quads per year (1.4 billion tons per year) by the year 2000, a 3 percent per year average growth rate (ERDA 1975, p. B-8). These two scenarios are combined—5 percent per year to 1985, slowing to 3 percent per year thereafter—as a reference or base case projection of coal demand.

Figure 6-20 illustrates the behavior of the coal sector when driven with the combined FEA-ERDA base case projection of coal demand. Because of the higher costs of underground production, almost the entire increase in coal production is satisfied by surface mining. Surface-mined coal increases to over 75 percent of total coal production by the year 2000, up from its 1975 value of 55 percent.

In the reference projection, coal production manages to keep pace with demand over the transition period (the coal production-demand ratio remains equal to 1). As Figure 6-20 illustrates, a coal supply problem will not appear if demand grows at 5 percent per year or less. Supply can keep pace with demand with no changes in the normal operations of the coal market.

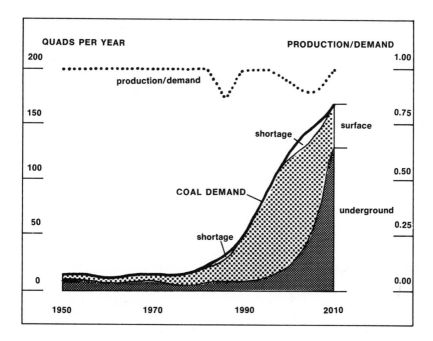

Figure 6-21. Accelerated demand projection

Accelerated Coal Demand Projection

If a national program were organized to shift toward the use of coal, coal demand could accelerate to much higher levels than those projected in the reference case. The FEA electrification scenario projects that coal demand could be increased to 29 quads per year (1.3 billion tons per year) by 1985, a 7.2 percent per year growth rate (FEA 1976, p. 185). The highest coal projection in the ERDA plan is 50 quads per year in the year 2000 (Scenario II in ERDA 1975, p. B-8). With policies that accelerate coal demand from both utilities and synthetic conversion plants, the COAL2 model projects that coal demand could grow as much as 7 percent per year over the long term, reaching almost 170 quads per year (7 billion tons per year) under extreme conditions. Coal demand is therefore assumed to grow at 7 percent per year in the accelerated coal demand projection.

Figure 6-21 shows the behavior of the coal sector when coal demand is accelerated to 7 percent per year to 2010. With accelerated demand, two distinct periods of coal shortages are forecast: (1) a startup problem from 1980 to 1990 as coal demand first begins to grow rapidly, and (2) a surface coal depletion problem from 2000 to 2010 when coal has grown to major importance in the United States energy system.

Coal Startup Problem. The first coal shortage is caused by the difficulties of moving the coal industry from stagnation to growth. To meet the rapid increases in demand forecast for the 1980s, the coal industry must invest in new capacity in the 1970s. Relative to normal coal industry standards, the investments needed during the 1970s are large, and coal prices must remain sufficiently high to attract adequate external financing (debt and equity) by establishing an attractive, stable return on investment. Yet such abnormally high coal prices are generated only from high coal demands, which fail to materialize until the 1980s, 5 to 10 years *after* the investment is needed.

The 1980-1990 coal shortage in Figure 6-21 is caused by a lack of financing. The financial needs of the coal industry are not met because of the industry's inability to generate funds from either internal or external sources. Two financial policies which might alleviate this problem—federal loan guarantees and coal price supports—will be tested later in this chapter.

Surface Coal Depletion Problem. As the coal industry settles into a high growth mode in the accelerated demand projection (after 1990 in Figure 6-21), financing of new mine capacity catches up with coal demand. The unit cost advantage of surface mines channels most of the new mine capacity into surface mines from 1975 to 1990, increasing surface-mined coal to 80 percent of total coal production by 1990.

Production of coal from surface mines reaches about 40 quads per year (1.6 billion tons), high enough to exhaust most of the premium, low-cost surface coal resources by 1990. As surface mining moves to thinner and deeper seams, surface production costs rise and new mine investment shifts heavily to underground mines. But the underground coal industry is in no position to provide additional coal when surface-mining operations begin to decline. Production of coal from underground mines is not much higher in 1990 than in 1970, since most of the early capital investment is in surface mining. As occurred in 1980 with the surface mining industry, the underground coal industry experiences its own startup problem. During the years 2000 to 2010 not enough new underground mine capacity can be brought on line to compensate for the drop in surface coal production.

By the year 2000, most of the 45 billion tons of premium surface-mineable coal resources are depleted. Production from surface mines shifts from a maximum of 85 percent of total production in 1995 to about 25 percent by the year 2010. Figure 6-21 indicates that policies restricting the use of surface coal and encouraging the development and use of the more abundant underground coal resources are needed to avoid long-term coal shortages. Further policy runs in this chapter will test the effects of surface-mining restrictions.

Financial Policies

The acceleration of coal production from its current stagnant levels to a 7 percent per year growth rate as called for by the accelerated coal demand scenario will require massive amounts of new capital investment well in advance of increased demand. The coal industry is incapable of generating enough investment funds to meet such a rapid increase in demand. To avoid the short-term (1980-1990) coal shortage of Figure 6-21, some sort of financial support is necessary.

Although a number of alternative financial plans could be devised to increase coal investment in the 1970s, only the two most commonly discussed government financial policies are tested here: debt guarantees and price supports. Debt guarantees tend to increase the industry's ability to attract external financing, while price supports tend to increase the attractiveness of new coal investments by stabilizing the rate of return. Each policy is examined for its effectiveness in alleviating the financial difficulties of the coal industry under high demand growth conditions.

Debt Guarantees. Eli Goldston and other coal industry spokesmen have called for government guarantees on coal company debt:

> The coal producers' basic problem . . . is to attain stability in earnings by way of stability in labor relations. Having accomplished that, the coal industry could gradually expand through internal cash generation and sale of its debt and equity securities, but never at the rate most energy forecasters expect will be required. To accomplish that rate of growth we need debt guarantees by the federal government such as have been used to spur expansion in other parts of the energy industry.
>
> Goldston 1973

A guaranteed loan policy is modeled in COAL2 by increasing the maximum external financing capacity of the aggregate coal industry (Figure 6-22). The higher investment capacity is consistent with an increase in the maximum external/internal financing ratio from 60/40 to 70/30. This high degree of external financing is not unprecedented in the energy industry. A similar level was achieved in the electric utility industry from 1951-1955 and from 1969-1971 (Hass, Mitchell, and Stone 1974, p. 61). However, the current coal

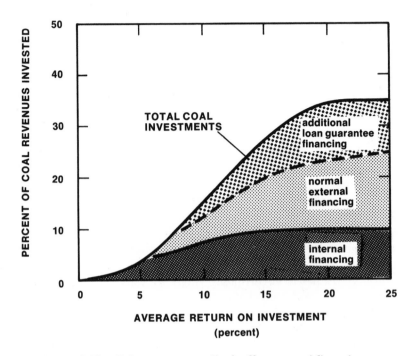

Figure 6-22.   Debt guarantee policy's effect on coal financing

industry external/internal financing ratio is 40/60 (MITRE 1974, p. 7-A-9). Therefore, a move to the maximum financing level of 70/30 with the debt guarantee policy represents a significant departure from current industry norms.

Figure 6-23 illustrates the behavior of the coal sector when a debt guarantee policy is added to the accelerated coal projection. The long-term forecast is not changed significantly with the addition of debt guarantees. The high rate of growth in coal demand is not met by coal production, and both the short-term shortage and the long-term shortage persist. Although the industry's debt capacity is expanded after 1977 in Figure 6-23, the industry does not utilize it fully. The coal industry has excess production capacity during the 1970s, because coal demand remains low. Profits and return on investment also remain low during this period, a market indication that new investments are not justified by demand.

When demand begins to rise after 1980, the coal industry can respond with greater external financing due to the debt guarantees. Therefore, the coal shortages are of shorter duration in Figure 6-23 than in the accelerated coal projection (Figure 6-21). But a debt guarantee policy does not anticipate the investment needs of the coal

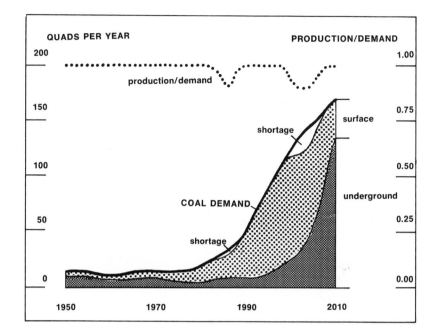

Figure 6-23. Debt guarantee projection

industry, and consequently is not an effective solution to the short-term coal financing problem.

Price Supports. Another method of government financial assistance for the coal industry is a coal price support. This policy is often discussed in connection with Project Independence goals (Hass, Mitchell, and Stone 1974, p. 98). The hypothesis is that, left to itself, the coal industry cannot attract enough capital because the projected demand for coal might never materialize if the price of imported oil dropped again. With price supports, coal companies would be guaranteed a minimum rate of return on their investment.

Figure 6-24 illustrates the behavior of the coal sector if coal prices are set to guarantee at least a 20 percent per year rate of return after 1977. Coal prices increase by 15 percent to an average of 45 cents per million Btu (11 1970 dollars per ton delivered), thereby stimulating higher investments in the late 1970s. When coal demand begins to rise after 1980, a sufficient amount of excess mine capacity exists to allow production to meet demand, as shown by the coal production-demand ratio continuing at 1.0 through 1990. The price support policy is effective in alleviating the short-term shortage because the policy *anticipates* the rise in demand, providing early

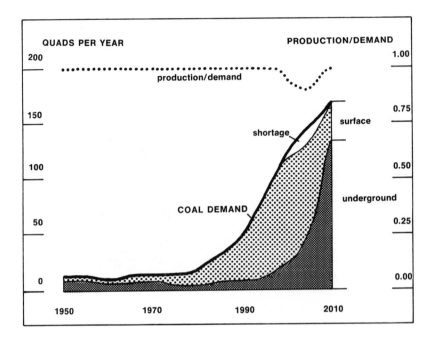

Figure 6-24. Price support projection

investment incentives. Neither the normal coal market response (Figure 6-21) nor the market aided by the debt guarantee policy (Figure 6-23) can meet the financial needs of accelerated demand because industry financing tends to respond to, rather than anticipate, changes in demand.

While a price support policy is effective in solving the short-term financial startup problems of the coal industry, coal production is still unable to meet demand from 2000 to 2010 due to the depletion of surface coal resources. The following section tests policies that affect the mix of surface and underground production over the long term.

Surface-Underground Policies

The ability of the coal industry to increase production depends not only on its financial viability but also on the allocation of investment between underground and surface mining methods. 1973 coal industry data indicate that underground coal production requires twice as much capital and two years more lead time than surface coal, at almost double the costs (NPC 1973, pp. 38, 139; FEA 1976, p. 199). Although surface coal mining holds a distinct advantage at the present, depletion of surface resources will shift the burden of coal production to underground mines during the next 30 years. Yet if the underground coal industry continues its decline over the next few decades, underground mines may be unable to meet national coal needs when surface resources are depleted.

Environmental factors will play a large role in determining whether surface or underground coal will dominate the expansion plans of the coal industry. Underground mining has historically been the most dangerous major American occupation (NSC 1962), due both to the risk of accidents and to the possibility of black lung disease. Continued high accident rates could constrain the development of the underground coal industry by contributing to a future dearth of coal miners. The Coal Mine Health and Safety Act of 1969 has imposed restrictions on underground mining practices and succeeded in lowering the number of mine fatalities and lost-time injuries. The effects of the 1969 act on labor availability and coal production will be examined in this section.

The major environmental issue associated with surface mining is reclamation. To date, surface mining has not been strictly regulated. State reclamation laws are largely unenforced, and in any case are most often designed to accommodate current mining practices (CRS 1973, p. 25). For example, in spite of the fact that Kentucky regulations are considered to be model state surface mining legislation (Train 1971), 40-50 percent of all state mine inspections reveal one or more violations (CRS 1973, p. 25). Two types of federal surface mining legislation are currently under consideration to eliminate environmentally unacceptable surface mining practices:

- partial or total ban on surface mining
- surface mining restrictions, including thorough reclamation (drainage, grading, revegetation), a ban on steep-slope mining (where effective reclamation is impossible), or perhaps even a tax on surface mining

Both policies will be tested to evaluate their impact on the long-term behavior of the coal industry.

1969 Health and Safety Act. The Coal Mine Health and Safety Act of 1969 has imposed a variety of restrictions on mining procedures in an attempt to insure the health and safety of the nation's coal miners. Although 1973 industry safety statistics following passage of the Act show fatalities halved, nonfatal injuries remain as high as before 1973 (*Coal Facts* 1974, p. 89). This analysis assumes that both fatal and nonfatal injuries will be significantly reduced as new procedures are fully put into effect.

The changes in operating procedures imposed by the 1969 act are assumed to reduce the productivity of underground mines by 25

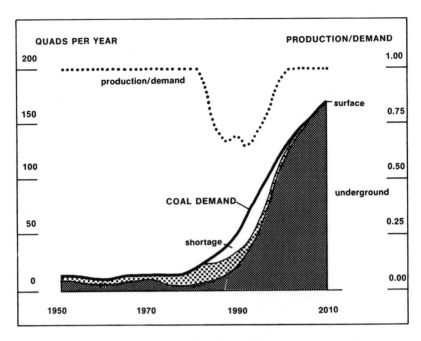

Figure 6-25.   No 1969 Health and Safety Act projection

percent (Conoco 1973, p. 19; CRS 1973, p. 24). To assess how much these new safety regulations have affected production, costs, and prices over the long term, the coal sector model is run *without* the 1969 act in Figure 6-25. This simulation should be compared to the Accelerated Coal projection (Figure 6-21) which includes the effects of the act in the model structure.

The immediate effect of the Health and Safety Act is twofold. Underground coal costs are increased by about 30 percent with the act, increasing average coal prices by about 15 percent. Underground production is slightly lower after 1970 due to decreases in productivity. Because the act increases underground mining costs, more investment flows to surface production. Coal production from surface mines is therefore higher over the long term with the 1969 legislation than without it.

Surface coal mines require less capital investment and can be completed in 3 years (rather than 5 years for an underground mine), and therefore more production capacity is acquired more quickly if investment is channeled towards surface mines during periods of increased demand. The short-term financial problems of the coal industry improve in Figure 6-21 when compared to Figure 6-25, which includes no Health and Safety Act. The improvement is manifested in the behavior of the coal production-demand ratio (the measure of a coal shortage), which indicates a less severe and more quickly resolved short-term shortage following passage of the act.

A comparison of Figure 6-21 and Figure 6-25 shows that the long-term shortage due to surface coal depletion is slightly improved by the 1969 act. Although the increased investment in surface mines worsens the severity of the surface coal depletion problem, the increased pool of underground miners resulting from improved safety conditions tends to improve the responsiveness of underground production. While the previous simulation has examined a policy already effected, the following runs look at new policies that might eliminate the long-term shortage.

Ban on All Surface Mining. In the past 15 or 20 years several states have passed laws and adopted regulations to control surface mining (CRS 1975, p. 11). But dissatisfaction with the lack of uniformity and effectiveness of current reclamation procedures has spurred numerous efforts by Congress to pass a national surface mine reclamation law. The law eventually passed could take one of two

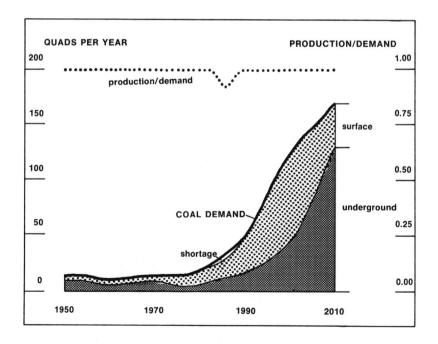

Figure 6-26. Ban on new surface mining projection

forms: (1) an outright ban on surface mining, or (2) restrictions on surface mining practices.

Figure 6-26 shows the effects of a phased ban on all new surface mining. All new coal investments are channeled toward underground mines after 1977, while existing surface mines are allowed to operate until their 20-year life is exhausted. The higher capital-output ratio of underground mining yields less mine production capacity per dollar of investment, and therefore the projected coal shortage is greatly amplified with a surface ban. Figure 6-26 projects a 20-year shortage of coal from 1980 to 2000. Given no additional financing policies, a ban on surface mining is incompatible with the accelerated coal demand schedules necessary for a massive transition to coal over the next 30 years.

Surface Mining Restrictions. A more likely and less severe form of federal surface mining legislation is enforcement of proper reclamation procedures and a ban on surface mining where proper reclamation is not possible. The requirement of thorough reclamation—drainage, grading, and revegetation—has been estimated to cost 20 cents per ton for Western coal, 1-2 dollars per ton for Eastern coal

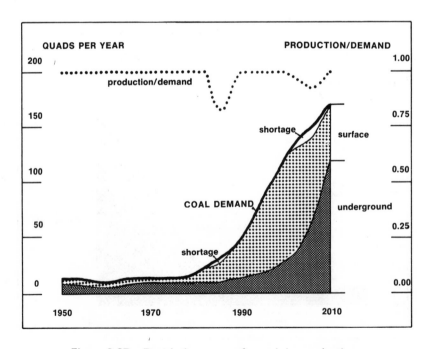

Figure 6-27.   Restrictions on surface mining projection

under moderate slopes, and up to 4.85 dollars per ton on the steep slopes typical of Appalachia (CEQ 1973, pp. 28, 29). A conservative estimate of 2 dollars per ton (8 cents per million Btu) is added to the cost of surface mined coal in COAL2 in order to simulate the effects of strict reclamation legislation.

The CEQ study on surface mining has also estimated the effect of a ban on mining where slopes are so steep that proper reclamation is impossible (CEQ 1973, p. 4). Using CEQ data, 3 billion tons of steep slope coal are removed from the COAL2 surface coal resource base to simulate steep slope restrictions. In addition, the CEQ study estimates that an immediate prohibition of surface mining on slopes of 15 degrees or greater would eliminate 7 to 18 percent of United States coal production (CEQ 1973, p. 6). Instead of an immediate ban, the COAL2 structure assumes that steep-slope mining is phased out gradually by prohibiting investment in new steep-slope mines. Consequently, steep-slope production capacity is lost gradually over the 20-year life of the remaining mines.

The economic impact of a strict federal reclamation law is relatively small, for even with the additional 8 cents per million Btu reclamation cost, surface-mined coal retains a distinct cost advantage over underground coal, approximately 35 versus 45 cents per million Btu. An additional tax of 12 cents per million Btu on surface-mined coal is added to further balance the development of surface and underground production which, as shown in Figure 6-21, is at present heavily skewed toward surface mining. The tax could be imposed regionally (a severance tax; see FEA 1976, p. 190) or federally. Federal revenues generated from the tax could well be used for developing ultimate energy sources such as solar or fusion.

Figure 6-27 shows the effect of these restrictive surface coal mining policies on the behavior of the coal sector. The reclamation restrictions and surface tax increase the cost of surface-mined coal to about 45 cents per million Btu in 1977, close to the cost of underground coal. Because underground coal is now cost-competitive, more investment flows to underground mines, resulting in a steady growth in underground coal production after 1977. Because the underground industry is thriving when surface coal production peaks and declines (near the year 2000), the transition to major dependence on underground coal proceeds smoothly. Only the short-term coal shortage brought about the financial limitations of the coal industry remains in Figure 6-27.

Combined Price Supports and Surface Restrictions

Two policies—price supports for the short-term financing problem (Figure 6-24) and surface mining restrictions for the long-term depletion shortage (Figure 6-27)—were found effective. Figure 6-28 illustrates the behavior of the coal sector if these two policies are combined. With both policies, both problems are avoided.

The price support policy increases total coal investment from 1977 onward, which effectively avoids the capacity shortage of the 1980s at the cost of creating excess coal capacity in the late 1970s. The surface coal restrictions tend to channel more of the total investment toward underground mining, creating a better balance between surface and underground coal production. Now that the underground industry is healthy and growing, it is better able to satisfy coal demand as surface production begins to decline after the year 2000.

CONCLUSIONS

If the United States is to engineer a massive shift away from its dependence on oil imports and toward the use of domestic coal, the coal industry must be able to increase its production rapidly over the next 30 years. The exact magnitude and timing of the demand depends on the types of policies implemented in the demand, oil and gas, and electricity sectors of the energy system. Chapter Seven of

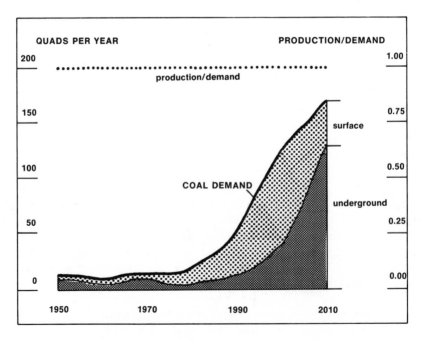

Figure 6-28.　Combination price support and surface restrictions projection

this volume considers the COAL2 model in its entirety, so that coal demand is generated as an endogenous variable. The accelerated coal demand scenario (7 percent per year growth in coal demand) provides an initial basis for examination of the expansion capability of the United States coal industry.

When the coal sector is driven by the accelerated coal demand scenario, two major periods of supply shortages result (Figure 6-21). As coal demand moves from its current status of no growth to high growth rates, a coal shortage occurs in 1980. The shortage persists for 10 years due to the inability of the industry to finance adequate new mine capacity. This short term startup problem can be alleviated by an anticipatory financial policy, such as coal price supports (Figure 6-24), which bolsters the financial status of the industry well before demand actually begins its rapid rise. A loan guarantee policy (Figure 6-23) proves a poor second choice in alleviating the short-term financial needs of the industry. Such a policy provides no incentives for early investment in new capacity.

The high rates of coal production in the accelerated coal projection create a second coal supply shortage near the end of the century (Figure 6-21). Because of its cost advantages, surface mining provides most of the increase in coal output from 1970 to 2000, while underground coal production, with its high capital, labor, and safety costs, stagnates near current levels. The rapid rise in surface production soon depletes the "premium" surface coal resources, causing surface production costs to rise rapidly and investment to shift to underground mines. The coal industry is faced in the year 2000 with a second financing problem. The underground coal industry cannot shift from stagnation to rapid increases in production capacity fast enough to avoid major capacity shortages. For ten years from 2000 to 2010 the industry experiences a difficult period where investment flows heavily to underground mining, yet the delays in opening new underground mines result in persistent undercapacity.

The current industry emphasis on surface mining is partly caused by the 1969 Coal Mine Health and Safety Act. To restore the balance between surface and underground production, similar restrictions could be placed on surface coal development. Although a total ban on surface mining (Figure 6-26) creates large short-term shortages, Figure 6-27 shows that less severe restrictive surface-mining policies, such as strict reclamation requirements, steep slope prohibitions, and a tax on surface-mined coal, are effective in alleviating the long-term coal shortage with little adverse effect on the short-term shortage. Surface mining restrictions tend to stimulate the underground coal industry towards growth in production, so that both surface and underground coal production increase from 1977 onward. When surface coal production peaks in 2000, the underground industry is capable of providing the needed additional mine capacity to satisfy total demand. Figure 6-28 combines price supports and restrictive surface policies to avoid both the short-term and the long-term coal capacity shortages associated with an accelerated coal development scenario.

The need for government intervention in the coal supply market is totally dependent on the dynamics of coal demand. Figure 6-20 establishes that low growth rates of coal demand (5 percent per year or less) would require no intervention in the coal supply system. With low growth in demand, the market is capable of "clearing" supply and demand with no severe capacity shortages. On the other hand, if coal demand does accelerate rapidly, federal policies will be necessary to stimulate enough production to meet demand. Analysis of the other three sectors of COAL2 (energy demand, oil and gas, electricity) can provide insights into the magnitude and timing of coal demand growth under a variety of energy policy options. Chapter Seven couples the coal sector with the other three sectors of COAL2, and examines both the need for and ability of coal to achieve the status of a major energy source during the energy transition period.

CHAPTER 6
EQUATIONS

```
100  *      COAL SECTOR OF COAL2
110  NOTE
120  NOTE   JANUARY 15, 1975
130  NOTE
140  NOTE   COAL SUPPLY-DEMAND BALANCE
150  NOTE
160  A      CD.K=DCUD.K+CDS.K+CDU.K+CED.K
170  A      CED.K=TABHL(CEDT,TIME.K,1950,2010,10)*1E15
180  T      CEDT=1.7/1.2/2/2/2/2/2
190  A      CCUF.K=TABHL(CCUFT,CD.K/CPC.K,0,1.2,.1)
200  T      CCUFT=0/.1/.2/.3/.4/.5/.6/.7/.8/.9/.95/.98/1
210  A      CPC.K=SCPC.K+UCPC.K
220  A      CPR.K=SCPR.K+UCPR.K
230  A      CPDR.K=CPR.K/CD.K
240  NOTE
250  NOTE   COAL FINANCING
260  NOTE
270  A      CPRICE.K=CPRAT.K*ACC.K
280  A      ACC.K=(SCPR.K*SCC.K+UCPR.K*UCC.K)/CPR.K
290  A      CPRAT.K=CLIP(CPRAT2.K,CPRAT1.K,TIME.K,PYEAR)
300  A      CPRAT1.K=TABHL(CPRAT1T,CD.K/CPC.K,.4,1.6,.2)
310  T      CPRAT1T=1/1.015/1.13/1.5/1.83/1.97/2
320  A      CPRAT2.K=TABHL(CPRAT2T,CD.K/CPC.K,.4,1.6,.2)
330  T      CPRAT2T=1/1.015/1.13/1.5/1.83/1.97/2
340  A      CROI.K=(CPRICE.K-ACC.K)*CPR.K/CC.K
350  A      CC.K=SC.K+UC.K
360  A      ACROI.K=SMOOTH(CROI.K,CROIAT)
370  C      CROIAT=5
380  A      FCRI.K=CLIP(FCRI2.K,FCRI1.K,TIME.K,PYEAR)
390  A      FCRI1.K=TABHL(FCRI1T,ACROI.K,0,.25,.05)
400  T      FCRI1T=0/.03/.13/.2/.23/.25
410  A      FCRI2.K=TABHL(FCRI2T,ACROI.K,0,.25,.05)
420  T      FCRI2T=0/.03/.13/.2/.23/.25
430  A      CREV.K=CPRICE.K*CPR.K
440  A      CCINV.K=FCRI.K*CREV.K
450  NOTE
460  NOTE   UNDERGROUND CAPITAL
470  NOTE
480  L      UC.K=UC.J+(DT)(UCICR.JK-UCDR.JK)
490  N      UC=UCI
500  C      UCI=2.5E9
510  R      UCDR.KL=UC.K/ALCC
520  C      ALCC=20
530  R      UCIR.KL=FIU.K*CCINV.K
540  R      UCICR.KL=DELAY3(UCIR.JK,UMCT)
550  C      UMCT=5
560  A      FIU.K=CLIP(FIU2.K,FIU1.K,TIME.K,PYEAR)
570  A      FIU1.K=TABHL(FIU1T,SCC.K/UCC.K,0,2.5,.5)
580  T      FIU1T=0/.3/.75/.95/.99/1
590  A      FIU2.K=TABHL(FIU2T,SCC.K/UCC.K,0,2.5,.5)
600  T      FIU2T=0/.3/.75/.95/.99/1
610  NOTE
620  NOTE   UNDERGROUND LABOR SUPPLY
630  NOTE
640  L      UCLS.K=UCLS.J+(DT)(NHR.JK)
```

```
650  N      UCLS=UCLSI
660  C      UCLSI=3.73E5
670  R      NHR.KL=CLIP(HR.K,LR.K,HR.K,0)
680  A      LR.K=(RCLS.K-UCLS.K)/LAT
690  C      LAT=.5
700  A      HR.K=(RCLS.K-UCLS.K)/HAT.K
710  A      RCLS.K=UC.K/RCLR.K
720  A      RCLR.K=TABLE(RCLRT,TIME.K,1950,2010,10)*1E4
730  T      RCLRT=.67/1.7/3.1/4.4/5.7/6.9/8.2
740  A      HAT.K=TABHL(HATT,PAR.K/(UW.K*HAWR),0,1.2,.2)
750  T      HATT=.5/.8/2/4/6.5/10/14
760  C      HAWR=.14E-3
770  A      UW.K=CLIP(UWP,UWH,TIME.K,PYEAR)
780  C      UWH=7000
790  C      UWP=7000
800  A      PAR.K=DLINF3(AR.K,SPDT)
810  C      SPDT=10
820  A      AR.K=SMOOTH(SS.K,SIDT)
830  C      SIDT=2
840  A      SS.K=CLIP(SSN.K,SSH,TIME.K,1970)
850  C      SSH=1
860  A      SSN.K=CLIP(SSP,SS69,TIME.K,PYEAR)
870  C      SS69=.5
880  C      SSP=.5
890  NOTE
900  NOTE   UNDERGROUND PRODUCTION
910  NOTE
920  A      UCPC.K=UCPC70*(LABR.K¨LE)*(CAPR.K¨CE)*PMS.K*UPMD.K
930  C      UCPC70=10.6E15
940  C      LE=.53
950  C      CE=.47
960  A      LABR.K=UCLS.K/UCLS70
970  C      UCLS70=108000
980  A      CAPR.K=UC.K/UC70
990  C      UC70=3.4E9
1000 A      PMS.K=TABHL(PMST,AR.K/ARH,0,1.4,.2)
1010 T      PMST=0/.41/.68/.82/.93/1/1.05/1.08
1020 C      ARH=1
1030 A      UPMD.K=TABLE(UPMDT,FUCRR.K,0,1,.2)
1040 T      UPMDT=0/.1/.22/.45/.9/1
1050 A      FUCRR.K=UCR.K/UCRI
1060 L      UCR.K=UCR.J+(DT)(-UCDPLR.JK)
1070 N      UCR=UCRI
1080 C      UCRI=15E18
1090 R      UCDPLR.KL=UCPR.K
1100 A      UCPR.K=CCUF.K*UCPC.K
1110 A      UCOC.K=UCAF*UC.K/(CCUFN*UCPC.K)
1120 C      UCAF=.65
1130 C      CCUFN=.8
1140 A      ULC.K=UW.K*UCLS.K/UCPR.K
1150 A      UCC.K=ULC.K+UCOC.K
1160 NOTE
1170 NOTE   SURFACE COAL PRODUCTION
1180 NOTE
1190 L      SC.K=SC.J+(DT)(SCICR.JK-SCDR.JK)
1200 N      SC=SCI
1210 C      SCI=1.4E9
1220 R      SCDR.KL=SC.K/ALCC
1230 R      SCIR.KL=(1-FIU.K)*CCINV.K
1240 R      SCICR.KL=DELAY3(SCIR.JK,SMCT)
1250 C      SMCT=3
1260 A      SCPC.K=SCCR.K*SC.K
1270 A      SCCR.K=SCCRN*SPMD.K
1280 C      SCCRN=3.5E6
1290 A      SPMD.K=TABLE(SPMDT,FSCRR.K,0,1,.2)
```

```
1300 T      SPMDT=0/.13/.35/.7/1/1
1310 A      FSCRR.K=SCR.K/SCRI
1320 L      SCR.K=SCR.J+(DT)(-SCDPLR.JK)
1330 N      SCR=SCRI
1340 C      SCRI=2.4E18
1350 R      SCDPLR.KL=SCPR.K
1360 A      SCPR.K=CCUF.K*SCPC.K
1370 A      SCC.K=SCAF/(SCCR.K*CCUFN)+SCP.K
1380 C      SCAF=.75
1390 A      SCP.K=CLIP(SCP2,SCP1,TIME.K,PYEAR)
1400 C      SCP1=0
1410 C      SCP2=0
1420 NOTE
1430 NOTE   EXOGENOUS INPUTS
1440 NOTE
1450 A      DCUD.K=TABLE(DCUDT,TIME.K,1950,2010,10)*1E15
1460 T      DCUDT=10.7/5.9/5.4/6/9/13/15
1470 A      CDS.K=TABLE(CDST,TIME.K,1950,2010,10)*1E15
1480 T      CDST=0/0/0/0/14/70/95
1490 A      CDU.K=TABLE(CDUT,TIME.K,1950,2010,10)*1E15
1500 T      CDUT=2.2/4.3/7.7/10/23/40/58
1510 NOTE
1520 NOTE   CONTROL CARDS
1530 NOTE
1540 A      FPS.K=SCPR.K/CPR.K
1550 N      TIME=1950
1560 C      PYEAR=1977
1570 SPEC   DT=.25/LENGTH=0/PLTPER=2/PRTPER=0
1580 PLOT   RCLS=R,UCLS=L(0,4E5)/CCUF=C,FPS=F(0,1)
1590 PLOT   CD=D,CPR=P,UCPR=U(0,2E17)/CPRICE=$(0,2E-6)/
1600 X      CPDR=*(0,1)
1610 RUN
1620 NOTE
1630 NOTE   PARAMETER CHANGES FOR THE COAL SECTOR RUNS
1640 NOTE
1650 NOTE   HISTORICAL RUN
1660 NOTE
1670 C      LENGTH=1975
1680 C      PLTPER=1
1700 PLOT   CD=D,CPR=P,UCPR=U(0,20E15)/CPDR=*(0,1)
1720 RUN    HISTORICAL RUN
1730 NOTE
1740 NOTE   BUSINESS AS USUAL COAL SCENARIO
1750 NOTE
1760 CP     LENGTH=2010
1770 T      CDUT=2.2/4.3/7.7/13/20/25/30
1780 T      CDST=0/0/0/0/0/0/0
1790 T      DCUDT=10.7/5.9/5.4/5/5/5/5
1810 PLOT   CD=D,CPR=P,UCPR=U(0,2E17)/CPDR=*(0,1)
1830 RUN    BUSINESS AS USUAL COAL SCENARIO
1840 NOTE
1850 NOTE   REFERENCE RUN: ACCELERATED COAL SCENARIO
1860 NOTE
1870 RUN    REFERENCE RUN
1880 NOTE
1890 NOTE   POLICY RUNS
1900 NOTE
1910 T      FCRI2T=0/.03/.15/.27/.34/.35
1920 RUN    DEBT GUARANTEES
1930 T      CPRAT2T=1.3/1.3/1.3/1.5/1.83/1.97/2
1940 RUN    COAL PRICE SUPPORT
1950 C      SS69=1
1960 C      SSP=1
1970 RUN    NO 1969 HEALTH AND SAFETY ACT
1980 T      FIU2T=1/1/1/1/1/1
```

```
1990 RUN    BAN ON SURFACE MINING
2000 C      SCP2=.2E-6
2010 T      SPMDT=0/.11/.32/.67/1/1
2020 RUN    SURFACE MINING RESTRICTIONS
2030 T      FIU2T=1/1/1/1/1/1
2040 T      FCRI2T=0/.03/.15/.27/.34/.35
2050 T      CPRAT2T=1.3/1.3/1.3/1.5/1.83/1.97/2
2060 RUN    SURFACE BAN & FINANCIAL POLICIES
2070 C      SCP2=.2E-6
2080 T      SPMDT=0/.11/.32/.67/1/1
2090 T      CPRAT2T=1.3/1.3/1.3/1.5/1.83/1.97/2
2100 RUN    PRICE SUPPORT & SURFACE RESTRICTIONS
```

Chapter 7

Energy Policy Analysis with COAL2

INTRODUCTION

In this chapter, the COAL2 model is used to analyze the effect of various energy policy options. The following sections summarize the objectives and methodological approach of the analysis, develop a *reference projection* of United States energy supply and demand, and, finally, show how the COAL2 model can be used to design an effective long-term energy program.

Problem Focus: Imports

Current energy policy is structured around the assumption that foreign oil imports will rise to fill any future gap between United States energy demand and domestic production. This assumption is built into the COAL2 model, and thus oil imports become the major indicator of an imbalance between United States energy supply and demand in the model output. Most of the ensuing analysis will focus on the future behavior of oil imports as projected by the COAL2 model.

Although the dominant energy policy goal is to shrink United States dependence on oil imports, both Congress and the Administration are equally concerned about the economic and environmental costs of achieving energy independence. The COAL2 model is very useful in evaluating potential side effects of Project Independence policies. The analyses in this chapter will focus on economic side effects. If the economic drawbacks of a proposed energy program are too great, the United States might be better off to accept high imports as a fact of life.

Basic Assumptions of the COAL2 Model

The COAL2 model is a dynamic computer simulation model composed of a complex, nonlinear feedback-loop structure. The precise assumptions that underlie its structure have been described in Chapters Two through Six. They can be summarized as follows.

Demand assumptions:

1. Energy demand is determined by *GNP* and the *efficiency* of energy use.
2. Energy price increases provide the incentive for improvements in end-use efficiencies.
3. Oil and gas, coal, and electricity are substitutable to some degree in the final energy markets. The degree of substitution is dependent on relative price of the fuels.
4. The consumer's response (efficiency changes, interfuel substitution) to a change in energy prices is delayed, as time is required to change consumption habits and technologies.

Supply assumptions:

1. Domestic production of energy is determined by the output capacity of production facilities, and the utilization of capacity.
2. Production capacity is dependent on the following factors:

conventional oil and gas — capital, resources
synthetic oil and gas — capital, R&D, coal availability
electricity — capital, environmental regulations, fuel availability
coal — capital, labor, resources, environmental and safety standards

3. The ability of the oil and gas, coal, and electric utility industries to generate new capital investment from internal (retained earnings) and external (debt, equity) sources is limited. Price regulation (in the oil and gas and electric utility industries) tends to reduce investments below the maximum limit.
4. There is a limited stock of recoverable oil, gas, and coal resources in the United States. As resources are depleted, the productivity of the capital stock decreases (the capital/output ratio increases).
5. Unavoidable delays limit the responsiveness of energy supply: for

example, 3- to 10-year construction delays, R&D delays (synthetic fuels, stack gas scrubbers), underground coal labor hiring delays, and response delays in the energy financing subsectors.
6. As oil and gas production capacity falls behind demand, the shortfall is made up with oil imports.

Figure 7-1 illustrates the basic structure of the COAL2 model, showing the major interconnections among variables. The diagram is divided according to the four sectors of the model: energy demand, oil and gas consumption, electricity generation, and coal production. Appendix B to this volume lists the DYNAMO equations for the COAL2 model and the parameter changes necessary to produce each of the simulation runs discussed in this chapter. Any reader with access to a computer equipped with a DYNAMO compiler can reproduce all the policy runs from the information in the appendix.

Using COAL2 for Policy Analysis

The current national debate over energy centers on whether or not government intervention is necessary in the United States energy system. Where should the United States place top priority: on demand-reducing policies or on policies designed to increase supply? Which policy options are most effective over the long term? To answer such questions, this chapter examines two demand programs (Accelerated Conservation and Zero Energy Growth), and two supply programs (Accelerated Nuclear and Accelerated Coal). Figure 7-2 furnishes a brief description of the policy changes included in each program and the way COAL2 model parameters were changed to simulate the program.

Six plots show how the model projects the effects of each policy. The first plot illustrates the behavior of *gross* energy demand and its supply components: domestic oil and gas, coal, nuclear, hydropower, and imports. This plot is of primary importance to energy policy-makers, for it provides a measure of future United States dependency on foreign oil. The availability of energy, measured by the energy consumption/demand ratio, is also shown in the first plot. If this ratio drops below 1.0, it indicates an energy shortage. Because energy demand is fuel-specific in COAL2, a shortage of energy could be caused by a shortage of any of the three energy forms: oil and gas, electricity, or coal.

The potential economic effects of a proposed policy are examined in the second output, which projects energy prices and the behavior of energy investments' share of GNP. Both the average price of energy paid by the consumer and the prices of the three final forms of energy—oil and gas, electricity and coal—are examined in this plot.

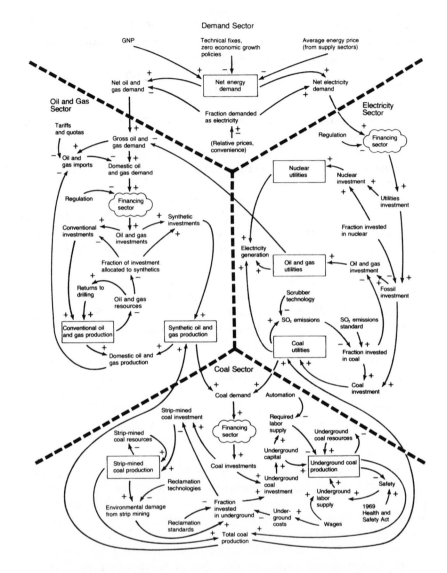

Figure 7-1. The structure of the COAL2 model

The energy investment/GNP ratio measures the potential macroeconomic effect of the energy investments generated by COAL2. Energy's share of fixed business investment for the total economy has averaged around 30 percent from 1947 to 1974 (FEA 1976, p. 295), while fixed business investment outlays have averaged about 10 percent of GNP during the same period (FEA 1974, p. 293). Energy investment/GNP has therefore averaged about 3 percent historically.

A Ford Energy Policy Project study on energy financing concludes that energy's share of fixed investment could *double* before the economy was adversely affected (Hass, Mitchell, and Stone 1974, p. 106). A separate study by Data Resources, Incorporated projects that fixed business investment could increase to 12 percent of GNP (FEA 1974, p. 293) with no major macroeconomic dislocations. Thus if energy investments/GNP rise above 7 percent in COAL2 (2 × 0.3 × 12), a capital shortage is indicated. Although the effects of such a shortage on the economy are not modeled in COAL2 , a high energy investment/GNP ratio serves as a warning signal in the model output.

The remaining four model plots outline the behavior of the four sectors of COAL2 : demand, oil and gas, electricity, and coal. The sector outputs are in the same format as presented in Chapters Three through Six for easy comparison of these whole-model runs with the previous sector-only runs.

COAL2 REFERENCE PROJECTION

The six plots comprising Figure 7-3 illustrate the *reference projection* of the COAL2 model. The figure presents the most likely evolution of the U.S. energy system *if no major changes in energy policy are enacted.* A detailed discussion of the "validity" of this reference projection (including the model's fit with historical behavior and the results of sensitivity tests) can be found in Appendix A.

Gross Energy Supply and Demand

Figure 7-3a illustrates the demand for and availability of gross energy inputs in the COAL2 reference projection. From 1950 to 1970, domestic oil and gas production satisfies most of the growing energy requirements of the United States, rising to over 75 percent of domestic energy consumption by 1970. Yet the model enters a totally different energy era after 1970. Domestic oil and gas production peaks in the early 1970s and declines thereafter due to depletion of the limited oil and gas resource base. By 1985, domestic oil and gas production satisfies only 25 percent of total U.S. energy needs. Their decline creates an urgent need for alternative sources.

Of the available domestic energy sources, only nuclear power and coal have the potential to contribute significantly to U.S. energy supplies over the 35-year time horizon of the COAL2 model. Yet, because of social, environmental, and economic constraints, neither coal (including coal-based synthetics and coal-fired electricity generation) nor electricity generated by nuclear power are able to replace domestic oil and gas rapidly enough to avoid a massive increase in

Program Name	Figure Number	Policies Included	COAL2 Parameter Change
Demand Programs			
(1) Accelerated Conservation	7-4	Incentive programs encourage conservation and implementation of more efficient end-use technologies.	Long-term price elasticity of total demand is increased from −.28 to −.40.
(2) Zero Energy Growth	7-5	a) Accelerated Conservation policies are implemented.	Same as above.
		b) Lifestyle changes stabilize material-goods consumption by 2010.	GNP stabilizes at 2 trillion 1970 dollars in 2010 (2 percent per year *average* growth rate).
Supply Programs			
(3) Accelerated Nuclear Program	7-6	a) Increase subsidies of the nuclear fuel cycle; government stimulates uranium exploration, accelerates the breeder program.	Nuclear fuel costs are reduced 20 percent below the reference run. (FEA-NTF 1974, p. 3.1-8)
		b) Standardized plant designs streamline the siting, safety, and environmental impact review process.	Nuclear project completion time is reduced from 10 to 6 years (AEC 1974, p. 6)
(4) Accelerated Coal Program	7-7	a) Oil and gas prices regulated at a 15 percent rate of return; a constant dollar floor is maintained on import prices ($12.00 per barrel in 1975 dollars).	Prices are set to allow a 15 percent ROI; import prices remain constant at $2.00 per million Btu.
		b) Research, development and demonstration of synthetic fuel technologies is accelerated.	The synthetic research and development lag is reduced to 4 years, allowing commercialization by 1981.
		c) A program of price guarantees, subsidies, or tax incentives encourages industry investments in commercial synthetic facilities.	The oil and gas investment decision is biased in favor of synthetics.
		d) Utilities are granted rate relief.	The regulatory lag between cost changes and price adjustments is removed; the allowed rate of return is increased to 10 percent per year.

e) Load-management policies reduce the season-ality of demand.

The average capacity utilization factor is increased to 60 percent.

f) SO$_2$ scrubbers are fitted on all coal-fired plants by 1990; low-sulfur coal is used in the interim.

Emissions per plant are reduced by 90 percent by 1990; capital costs are increased $50.00 per kilowatt.

g) Coal price supports are implemented in 1977.

Coal companies are guaranteed a 20 percent rate of return on invest-ment.

h) Surface mining is banned on steep slopes (>15°).

Recoverable surface coal resources are decreased by 3 billion tons.

i) Strict federal reclamation requirements are enacted; a Btu tax is imposed on surface-mined coal.

Surface coal costs are increased by 20 cents per million Btu.

Figure 7-2. Policy options tested with COAL2

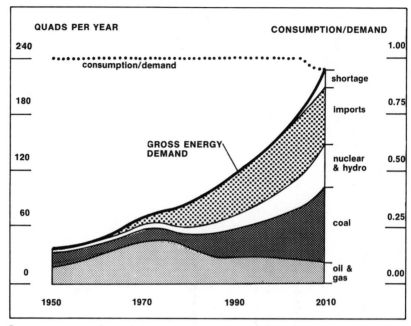

a. Gross energy supply and demand

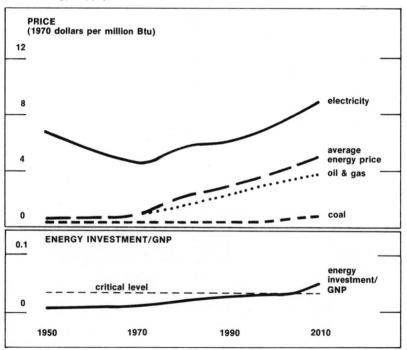

b. Economic effects

Figure 7-3.   COAL2 reference projection

foreign oil imports. In the COAL2 reference run oil imports increase by a factor of *three* from their current (1975) level to 36 quads per year (17 mbpd) in 1985. Imports increase to 37 percent of gross U.S. energy consumption in 1985, compared to the current 20 percent dependence level. More seriously, the import problem persists through the end of the century. By the year 2000, imports still supply almost 40 percent of U.S. energy inputs.

Figure 7-3b plots the behavior of the economic variables in the COAL2 reference projection. By 1985, real oil and gas prices increase bv a factor of two from 1975 levels to $2.00 (1970) per million Btu ($12.00 per barrel). By the year 2010 oil and gas prices reach $4.00 per million Btu, four times 1975 prices. Electricity prices also begin to rise after 1970, due to increasing capital costs and fuel prices. Coal prices are the most stable, rising only about 20 percent during the 35-year period. The average energy price paid by consumers increases by a factor of two by 1985. Because energy production becomes increasingly capital-intensive, the energy investment/GNP ratio steadily rises in Figure 7-3b. A shift to synthetic fuels by the oil industry precipitates a capital shortage after the year 2000 in the reference run.

Causal Mechanisms of the Energy Problem

Figures 7-3c through 7-3f illustrate the behavior of the four individual sectors of the energy system (demand, oil and gas, electricity, and coal) that interact to create the rapid deterioration of the U.S. energy supply-demand balance shown in Figure 7-3a. The variables and scales chosen for each sector plot are identical to the outputs shown for the sectors when run individually in Chapters Three through Six.

Energy Demand Sector. Figure 7-3c shows the behavior of the energy demand sector if no new energy policies are initiated. Although GNP is assumed to grow at 3.3 percent per year to the year 2000, increases in energy prices (Figure 7-3b) reduce the growth in *net* energy demand to about 2 percent per year from 1975 onward. The high price of oil and gas encourages a shift away from use of oil and gas in the final energy markets and toward electricity and coal, whose prices are rising more slowly. Although the fraction of net energy demanded as oil and gas decreases from 80 percent in 1975 to 50 percent in the year 2010, the total amount of oil and gas demanded continues to increase through the next 35 years (though at a much slower rate than the past quarter-century). Even though coal prices are the most stable of the three energy alternatives (coal, oil and gas, and electricity) over the long term, direct demand for coal remains a small fraction of net energy demand over the transition period. Due to delays in designing and constructing new coal-burning devices, and to the strong consumer preference for oil, gas, and electricity, direct coal use is limited primarily to industry (see Chapter Three).

The trend toward electricity is responsible for the increasing disparity between gross and net energy demand through the year 2010. Only about one-third of the energy content of the input to a steam-electric utility is converted to electricity. Therefore, increasing the use of electricity tends to decrease the overall efficiency of the energy system. While less than 20 percent of the gross energy inputs were lost in the conversion of one form of energy to another in 1975, conversion losses increase to 45 percent by 2010. The increased use of electricity could reduce the demand for imports if the additional electricity were produced from sources other than oil and gas. But, in the reference run, the trend toward electricity use actually exacerbates the transition problem.

Oil and Gas Sector.   Figure 7-3d illustrates the behavior of the oil and gas sector when the COAL2 model is run with no change in existing energy policies. Oil and gas demand rises at its historical rate of 5 percent per year from 1950 to 1970, and then slows its rate of growth considerably (to less than 2 percent per year) through the year 1990 due to the rapid rise in oil and gas prices in the 1970s. Yet, simply slowing the rate of growth of oil and gas demand is not enough to stem the rise in imports. Domestic oil and gas production from conventional wells declines rapidly after 1972. The peak and decline of domestic production, coupled with continued increases in demand, creates a sudden surge of imports after 1970. Imports nearly double from 7 quads per year in 1970 to 13 quads per year in 1975. By 1985 U.S. oil and gas imports increase to 36 quads per year (17 mbpd), over 50 percent of total oil and gas consumption.

Figure 7-3d shows clearly that the bell-shaped behavior of oil and gas production is a major contributor to the rapid rise in U.S. imports. The primary cause of the post-1970 decline is depletion of the finite oil and gas resource base. Total recoverable oil and gas resources are estimated at 2,800 quadrillion Btu (480 billion barrels of oil equivalent; see Chapter Four). By 1975, 1250 quadrillion Btu (215 billion barrels of oil equivalent), representing 45 percent of the oil and gas resource base, have been consumed, leaving only the more expensive resources in less accessible locations (small pools, offshore, Alaska) for future production. Production declines as less oil and gas is extracted per unit of capital investment, or per foot of wells drilled.

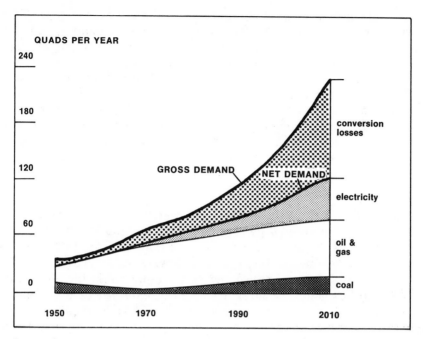

c. Demand sector

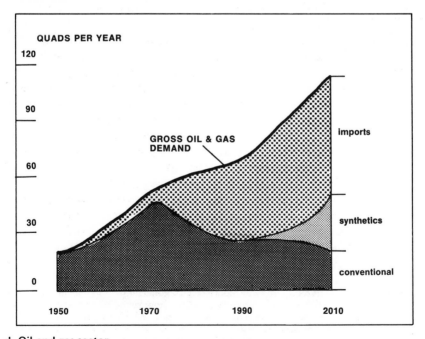

d. Oil and gas sector

Figure 7-3 (continued). COAL2 reference projection

If oil and gas demand continues to rise as in Figure 7-3d, producers of oil and gas have three sources of supply: imports, production from conventional wells, and synthetic oil and gas conversion facilities. By 1975 all these sources are expensive: the price of new conventional oil and gas and of oil imports is nearly $15.00 per barrel in 1975 dollars (Nathan 1975), and it is now estimated that the price of synthetics will be $21.00-24.00 per barrel (Hammond and Zimmerman 1975, p.46). Due to a lack of investment incentives, investment in synthetic fuels does not begin until after 2000. By default, imports become the primary U.S. energy transition fuel.

Electricity Sector. In Figure 7-3e electricity demand grows at its historical rate of about 7 percent per year from 1950 to 1973, during a period of declining electricity prices. From 1975 to 1990 electricity demand grows at 6 percent per year. The continued high electricity growth rates are caused by the rapid rise in the price of its substitute—oil and gas—after 1973. Electricity prices continue to decline relative to oil and gas prices after 1970, encouraging substitution of electricity for scarce and expensive oil and gas.[a] After 1990, electricity demand slows its growth to 4.5 percent per year (as oil and gas prices grow more slowly) due to the reductions in the growth rate of net energy demand.

When the electricity sector was run in Chapter Five, the individual sector behavior exhibited the potential for both a capacity shortage and a heavy dependence on oil and gas as a boiler fuel. Both problems occur in Figure 7-3e when the COAL2 model is run as an integrated whole. The capacity shortage is caused by a decline in the average rate of return of electric utilities after 1970, as utility rates fail to keep pace with rising costs. Although nuclear power grows significantly in the COAL2 reference run (to over 100,000 megawatts by 1985), it generates only 32 percent of U.S. electricity needs—and 11 percent of total net energy needs—by the year 2010.

Nuclear power is blocked from making a major contribution to total U.S. energy supplies by two factors. First, nuclear power generated only about 8 percent of U.S. electricity needs in 1975. Even if nuclear power grew at the more optimistic rates projected by

---

[a]An increase in oil and gas prices also causes electricity prices to rise, for fuel costs are an important part of the consumer's electricity rates. Yet only approximately 35 percent of U.S. electricity is generated from oil and gas (EEI 1974a, p.22). Furthermore, before the 1973 increase in oil prices, fuel costs comprised only about 25 percent of the price of electricity (EEI 1974a, pp.48,53). Thus the large increase in oil and gas prices shown in Figure 7-3b accounts for only about 20 percent of the electricity price increase. The other 80 percent of the electricity price increase is attributable to rising capital costs.

the AEC (AEC 1974, p. 6; ERDA 1975, p. B-11), nuclear's share of the electricity market would increase to only 50 percent by the year 2000 (see Chapter Five). Second, electricity currently satisfies only 10 percent of net energy demand. Even with the marked increase in electricity's share shown in the reference run (Figure 7-3c), nuclear power is capable of satisfying a maximum of only 20 percent of net energy demand by the year 2010, and much less during most of the transition period. *Under any circumstances, the U.S. must depend on fossil fuels for most of its energy (and even for most of its electricity) during the next 35 years.*

Due to the rapid rise in oil and gas prices after 1973, most of the additions to fossil-fired utility capacity during the 1970s consist of plants designed to burn low-sulfur coal. But total SO_2 emissions soon grow to intolerable levels even with the use of low-sulfur coal. Investment in coal-fired utilities drops substantially after 1985 in response to pressure to comply with environmental standards. The electric utility industry then shifts to heavy reliance on oil and gas as a boiler fuel, causing oil and gas demand from utilities to increase sharply after 1990 in Figure 7-3c.

Coal Sector. Because of the lack of incentives for development of synthetic fuels from coal and the continued environmental constraints on using coal in electric utilities, coal demand rises at only 4 percent per year from 1975 to 2000 in the reference run (Figure 7-3f). Demand increases from 15 quads per year (620 million tons) in 1975 to 24 quads per year (1 billion tons) in 1985, equal to the FEA reference scenario for coal development (FEA 1976, p. 33). The coal industry satisfies this demand without government interference in the COAL2 model. Due to its cost advantage, surface mining provides most of the needed new mine capacity, while production from deep mines continues near current levels.

For 25 years (1975-2000), the coal industry maintains stable, modest growth. Yet after 2000, a major coal shortage appears. The oil and gas industry has begun to build synthetic fuel plants (Figure 7-3d), which accelerates the growth in coal demand to 9 percent per year. Because the coal industry has made a major shift to surface coal production, most of the 45 billion tons of premium surface coal resources have been depleted by the year 2000. In the reference run, the underground coal industry is unable to meet the rapid increases in demand from synthetic conversion plants.

Reference Projection Summary

The COAL2 reference projection is substantially more pessimistic in its forecast of the current direction of the U.S. energy system than

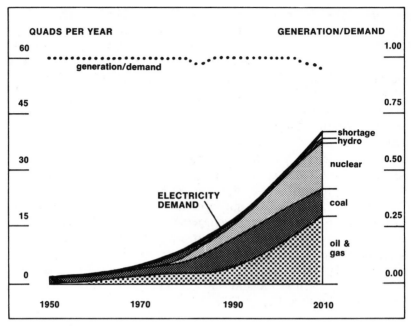

e. Electricity sector

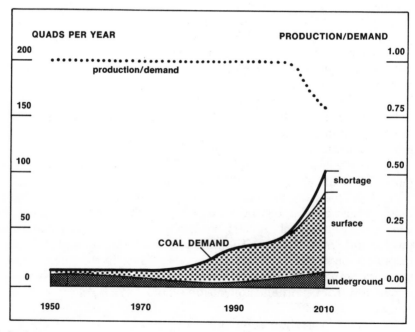

f. Coal sector

Figure 7-3 (continued). COAL2 reference projection

increased incentives are modeled in COAL2 as influences on the oil and gas investment decision. Due to the increased incentives, financial decisionmakers in the oil and gas industry shift more investment into commercial synthetic facilities than is justified by normal financial considerations (cost and rate of return estimates).

The relationship controlling investment in commercial synthetic plants in COAL2 is shown in Figure 7-4. The reference run assumes that investment in synthetic facilities would begin in earnest only when synthetic and conventional oil and gas production costs reach parity (conventional/synthetic cost ratio equals 1). As a result of the Synthetic Fuels Commercialization Program, the investment decision would be biased toward synthetics. The relationship shown in Figure 7-4 is shifted upward to higher synthetic investments at each cost comparison. This policy change is part of the Accelerated Coal Program shown in Figure 7-2 and tested later in this chapter. With this change in place, the COAL2 model can map out a new behavior of the system, including the direct effects on synthetic fuels investment and production, and all the indirect effects on conventional oil and gas production, imports, electricity generation, coal production, and energy demand.

Because the model focuses on the long-term behavior of the energy system, many effects outside the model's boundary (such as inflation, GNP growth, or unemployment) will not show up in the output. For this reason, the results of the COAL2 model policy analysis must be considered in conjunction with analyses that explicitly examine other aspects of a proposed policy change.

Energy Demand Policies

Future growth in energy demand will depend on two factors: (1) *growth* in the use of energy-consuming goods and services, and (2) changes in the *efficiency* of energy use by the expanding output of goods and services. In recognition of this distinction, the Ford Foundation Energy Policy Project (Ford 1974a) has suggested that energy demand policies naturally fall into two categories: policies that encourage greater conservation and end-use efficiency (Ford calls these Technical Fix policies; we use the term Accelerated Conservation) and policies that stabilize the final demand for energy (Zero Energy Growth policies). The effectiveness of each of these demand-reducing strategies has been tested with the COAL2 model.

The reference run already embodies a substantial amount of demand reduction resulting from increasing energy prices. The Accelerated Conservation policy package suggests a number of

the projections of most other energy analyses. If no major policy changes are implemented, the COAL2 model projects an increase in imports by a factor of *three* to 36 quads per year (17 mbpd) by 1985. In contrast, the ERDA Plan (ERDA 1975, p. V-2) projects a "no new incentives" imports level of 26 quads per year and the Federal Energy Administration projects 28 quads per year at a world oil price of $13.00 per barrel (FEA 1976, p.xxvii).

The discrepancy in projections of future imports stems directly from the basic assumptions of each analysis. The major cause of the rapid increase in imports in the COAL2 model is the inclusion of two factors: depletion and delays. The COAL2 model includes the effects of depletion of the finite oil and gas resource base, which causes oil and gas production to drop substantially after 1970. Although both the FEA and ERDA analyses have estimated the effects of resource depletion, neither group has formally analyzed the long, unavoidable delays involved in reducing demand and developing alternative energy sources. The difference in imports between COAL2 and the other studies (36 quads versus 26-28 quads) can be attributed to the inclusion of response delays in COAL2.

In the COAL2 reference projection, the U.S. rapidly reaches a high level of vulnerability to undesirable external influences on U.S. foreign and domestic policy (almost 40 percent of U.S. energy is imported in 1985). Indeed, the projected increase in U.S. petroleum imports, when coupled with similar import rises in other Western economies, could deplete the seemingly abundant world crude oil resources before the end of the century (Hubbert 1969, p.196). Because of the massive, long-term dependence on foreign energy sources projected in the COAL2 reference run, major changes in national energy policy are indicated.

ENERGY POLICY ANALYSIS WITH COAL2

The most important feature of the COAL2 model is its ability to analyze the effects of changes in energy policy. Because the structure of COAL2 models the investment, production, and hiring policies that control the behavior of the energy system, it is relatively easy to change those model policies to match proposed real-world policy changes.

For example, the Administration recently proposed a program to provide government incentives for the development of synthetic fuels. The proposal is reported in detail in a four-volume interagency study, *Recommendations for a Synthetic Fuels Commercialization Program* (SITF 1975). The recommended program includes government-guaranteed loans and price supports for synthetic fuels. The

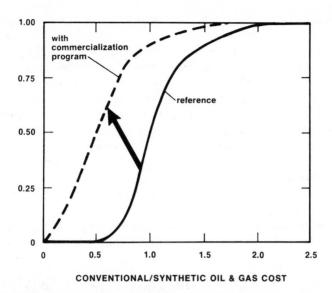

Figure 7-4. Shift in conventional/synthetic investment decision made to model a Synthetic Commercialization program

government, industry, and consumer actions to reduce final demand further. The Zero Energy Growth policy package, on the other hand, suggests fundamental changes in the patterns of production and consumption of final goods. These potential changes include substituting labor for energy in the production of goods, increasing product lifetimes, and reducing growth in final demand for energy-consuming goods.

Accelerated Conservation Projection. The Ford Foundation Energy Policy Project has specified a number of energy-conserving technologies which, if carried out, would bring direct energy savings. These technologies include the use of better insulation, heat pumps, more efficient heating and cooling units, total energy systems where energy producing and consuming units are integrated to avoid energy waste, improved automobile fuel economy, more efficient production processes, and the use of heat recuperators and regenerators (Ford 1974a, pp. 50, 52, 58, 64). Projected energy savings would be achieved by supplementing market forces with the following government policies (Ford 1974a, pp. 53, 62, 68):

- improve consumer information with full disclosure of energy costs

- legislate tax incentives for insulation retrofit of homes and commercial buildings
- revise building codes to be consistent with building economics based on life cycle costs at current (high) prices of energy
- eliminate gas pilot lights in appliances and equipment
- legislate fuel economy regulations for automobiles
- reform Interstate Commerce Commission regulations to encourage using railroads
- improve short-haul rail passenger service
- encourage the Civil Aeronautics Board to reform air traffic scheduling to increase load factors
- encourage the use of on-site power generation through differential rates
- remove the disincentives for use of recycled materials in the federal income tax structure and in railroad transportation rates

The long-term, aggregate effect of all these policies in the Accelerated Conservation program is modeled in COAL2 as an increase in the price elasticity of total demand. In response to energy price increases, more energy-efficient technologies become economical than in the reference run. The widespread use of such energy-saving technologies is projected to have a negligible effect on the economy's total production (Ford 1974a, p. 71). Consequently, in the COAL2 model, growth in GNP is assumed to be unaffected by the Accelerated Conservation policies. Gross energy consumption is reduced 4 percent below the reference run in 1985 to 94 quads per year, and 12 percent savings are realized by 2000. For comparison, the FEA projects that gross energy consumption will rise to 93 quads per year in 1985 with Accelerated Conservation policies (FEA 1976, p. G-5). Accelerated Conservation policies reduce the growth in gross energy demand to an average of 2.5 percent per year to the year 2000.

Although energy demand is reduced with Accelerated Conservation, Figure 7-5a shows that a significant domestic energy gap still persists. Imports remain near 35 quads per year, three times current levels, from 1985 to 2010. Accelerated Conservation policies do little to change the mechanisms responsible for the increase in energy imports. Demand still grows, depleting the finite oil and gas resource base. Significant impediments still block the development of alternative domestic energy sources—price regulation, technological delays, and environmental constraints. Reliance solely on technological advances in the demand sector to solve the energy problem would hardly dent the huge increase in U.S. dependence on imports projected for the next 35 years.

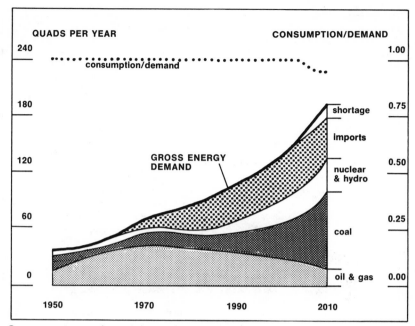

a. Gross energy supply and demand

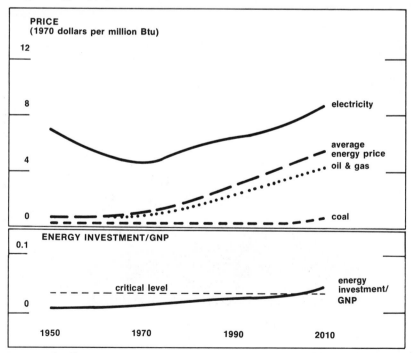

b. Economic effects

Figure 7-5. Accelerated Conservation projection

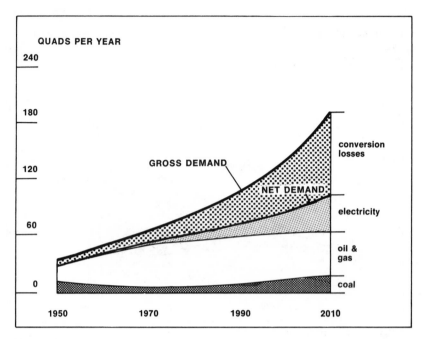

c. Demand sector

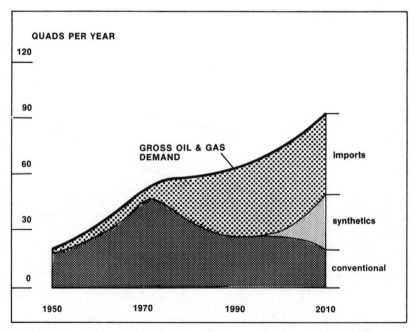

d. Oil and gas sector

Figure 7-5 (continued). Accelerated Conservation projection

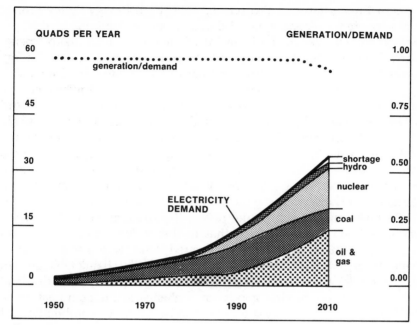

e. Electricity sector

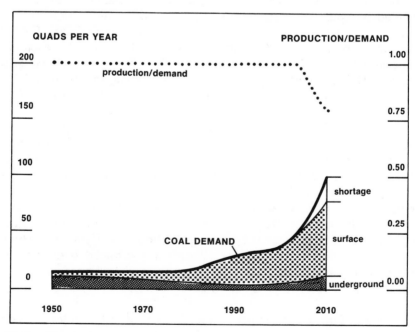

f. Coal sector

Figure 7-5 (continued). Accelerated Conservation projection

Zero Energy Growth Projection.   The Ford Energy Policy Project has suggested a second type of demand policy, implying gradual yet permanent changes in the behavior of energy demand: Zero Energy Growth (ZEG). Speculating about the implications of a Zero Energy Growth policy on the nature and path of economic development is an expedition into uncharted waters. The Ford Project claims, through analysis with the Data Resources econometric model (Ford 1974a, Appendix F), that energy and GNP growth can be effectively decoupled, allowing GNP to continue to increase as energy use stabilizes. This "decoupling" is accomplished by a major shift in the *composition* of GNP toward less energy-intensive products, super-imposed on an economy already adapted to the energy-efficient Accelerated Conservation policies (Ford 1974a, p. 509).

To make the COAL2 model reflect Zero Energy Growth policies, we have made the assumption that further reductions in the growth of energy demand below the Accelerated Conservation scenario must come from reductions in GNP growth itself. Stabilizing GNP growth in the model therefore implies a leveling off in the production and use of energy-consuming goods, or a change in the composition of GNP toward less energy-intensive goods, or some combination of these two.

If Zero Energy Growth were carried out gradually and accom-panied by continued reductions in population growth, high levels of material consumption per capita could be maintained. Indeed, gradual stabilization of GNP by 2010 would allow GNP to reach two and a half times its 1975 level by 2010. When considered with current U.S. population projections, GNP per capita would stabilize at 1.5 to 2 times current levels with the ZEG policies (Bureau of the Census 1972).

Figure 7-6 illustrates the effects of a Zero Energy Growth policy on COAL2 model behavior. To model ZEG, the Accelerated Conser-vation increases in demand elasticity are combined with a reduction in the GNP growth rate to zero by 2010. In the ZEG projection, net energy demand stabilizes at 75 quads per year by 2000—a 25 percent reduction from the reference projection. Gross energy demand is also substantially reduced (to 115 quads in 2000), yet continues to grow due to the shift to electricity and synthetic fuels. Figure 7-6a shows that the oil import problem is substantially reduced over the long term. With Zero Energy Growth, imports supply 35 percent of gross energy inputs in 1985, and only 5 percent in 2010.

Demand Policies Summary.   Most of the suggested policy options for reducing energy demand have focused on "conservation" incen-tives, where the efficiency of energy use is improved (CED 1974,

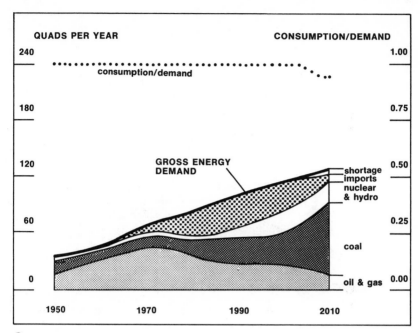

a. Gross energy supply and demand

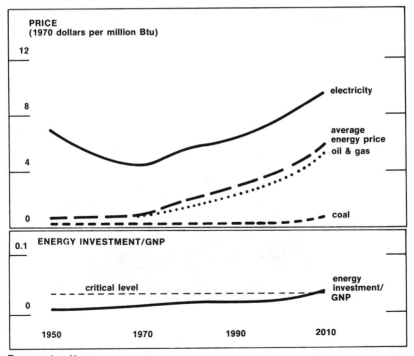

b. Economic effects

Figure 7-6.   Zero Energy Growth projection

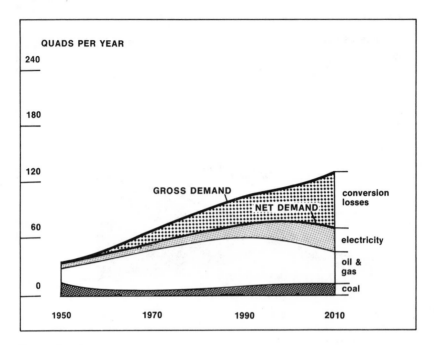

c. Demand sector

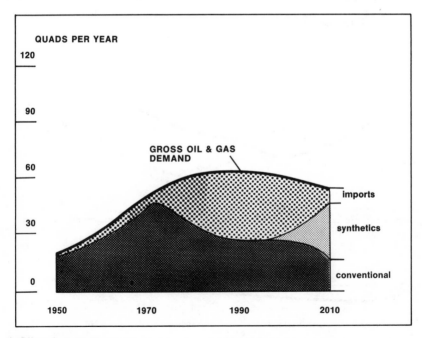

d. Oil and gas sector

Figure 7-6 (continued).   Zero Energy Growth projection

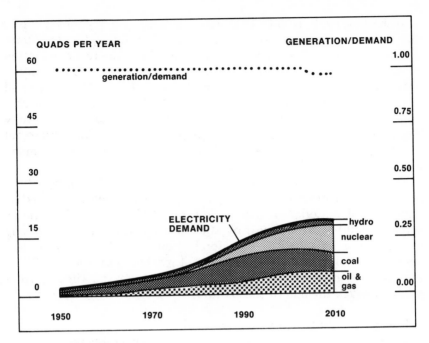

e. Electricity sector

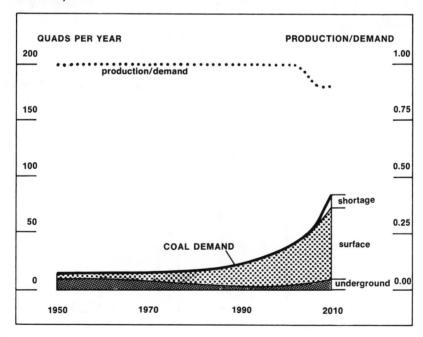

f. Coal sector

Figure 7-6 (continued).   Zero Energy Growth projection

p.30; ERDA 1975, p. 5-6; FEA 1976, p. E-6). Yet Figure 7-5 shows that improving energy efficiencies hardly modifies the long-term behavior of the U.S. energy system. When the 15 percent reduction in energy demand that results from the Accelerated Conservation policy is spread over the 35 years from 1975 to 2010, demand increases from GNP growth tend to overwhelm the gains in efficiency. Even if the Accelerated Conservation policies were fully implemented in 1975, continued growth in GNP would increase net energy demand back to previous levels in only 5 years, and demand would continue to grow at 3 percent per year thereafter.

The Zero Energy Growth policy tested in Figure 7-6 brings about much more profound changes in energy demand behavior. Gradual reduction of material growth to zero by 2010 stabilizes net energy demand at 90 quads per year by 2000 and allows the U.S. eventually to achieve independence from foreign oil imports. Yet ZEG is a long-term strategy—independence is not reached until the year 2010. Imports remain above 30 quads per year (about 2.5 times current levels) through the end of the century.

Energy Supply Policies

A comparison of the nation's ultimately recoverable resource base with present U.S. consumption patterns makes the reason for the current energy imbalance abundantly clear. Figure 7-7 illustrates that 76 percent of the 1974 U.S. energy usage was dependent on oil and natural gas, our least plentiful energy resources. The most abundant resources, coal and uranium, remain relatively neglected. For example, the remaining recoverable resources of coal are almost 10 times

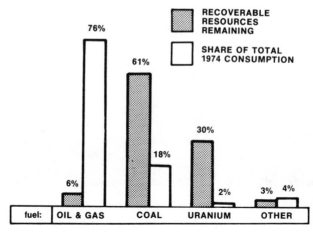

Source: FEA 1976, p. xxii.

Figure 7-7. Fuel availability vs. usage

those of petroleum, natural gas, and oil shale combined. The transition away from a major reliance on oil and gas is already long overdue. Oil and gas production peaked in the early 1970s, and is beginning its inevitable decline due to resource depletion effects. Because of the high price of the domestic alternatives, coal and nuclear power, government incentives are necessary to shift U.S. dependence to either of these sources. This section tests the effects of two supply incentive programs, an Accelerated Nuclear program and an Accelerated Coal program, to see if a major dependence on imports could be avoided by stimulating domestic production.

Accelerated Nuclear Projection. The development of nuclear power is accelerated in COAL2 by changing two parameters. First, delays in constructing nuclear power plants are reduced from the current average of 10 years to 6 years. The reduced delay represents a streamlining of the siting and permit-reviewing process. According to the AEC "Case C":

> This case assumes additional improvements in construction performance and regulatory processes. New legislation and rules would permit construction to begin prior to completion of the construction permit application safety review. The site environmental review would be completely separated from the safety review. This presupposes that standardized plant designs would be used in the license application. The project time would be about 6 years with 1 year for design and planning, license application preparation and environmental review and 5 years for construction and start up with concurrent operating license review and approval.
>
> AEC 1974, p. 6

Second, nuclear fuel cost projections are reduced 20 percent below the reference run through increased federal subsidies for the nuclear fuel cycle and a concerted national program to stimulate exploration for new uranium reserves. As suggested in numerous AEC reports, such an accelerated program would have to include the timely introduction of the breeder reactor to dampen the effects of uranium depletion on nuclear fuel costs (FEA-NTF 1974, p. 2.0-7).

Figure 7-8 shows the behavior of COAL2 when the Accelerated Nuclear program outlined above is started in 1977. Nuclear power capacity grows to 160 gigawatts[b] in 1985, very close to the maximum FEA projection (FEA 1976, p. 38). However, even accelerated development of nuclear power does not significantly improve the U.S. energy balance. Because electricity from nuclear and coal-fired plants costs virtually the same after 1977, electricity

---

[b]One quad per year electrical output is approximately equal to 67 gigawatts capacity.

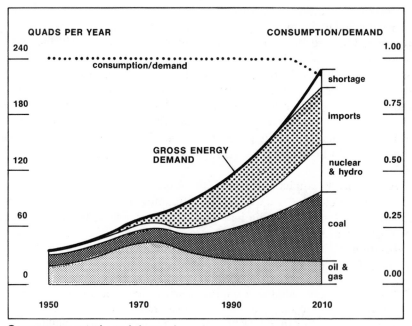

a. Gross energy supply and demand

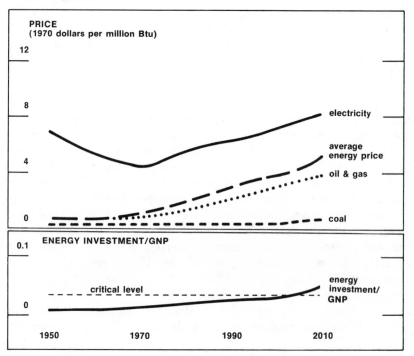

b. Economic effects

Figure 7-8.   Accelerated Nuclear projection

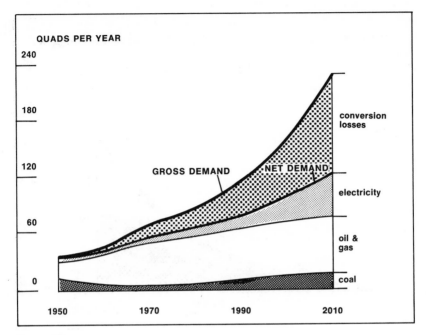

c. Demand sector

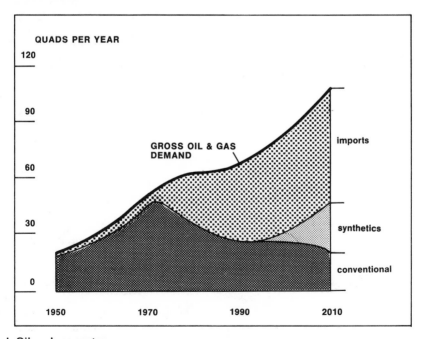

d. Oil and gas sector

Figure 7-8 (continued).   Accelerated Nuclear projection

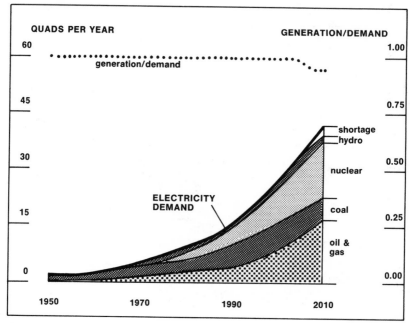

e. Electricity sector

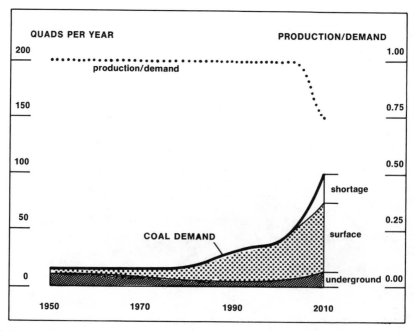

f. Coal sector

Figure 7-8 (continued).   Acclerated Nuclear projection

demand is not increased by the Accelerated Nuclear program. Most of the new utilities constructed after 1977 are either coal-fired or nuclear plants, and so the Accelerated Nuclear program tends to reduce future coal consumption in utilities, rather than provide a substitute for imports. United States dependence on oil imports drops only 10 percent below the reference run in the year 2000. Oil imports are still the dominant transition energy source, providing over 35 percent of gross energy inputs from 1985 onward.

Accelerated Coal Program. If coal is to replace oil and gas as the dominant fuel source during the transition period, substantial barriers to its production and use must be overcome. The current level of regulated oil and gas prices offers little incentive for development of coal-based synthetics, estimated to cost $20.00-24.00 per barrel (MIT 1974, p. 46). Construction of coal-fired utilities has been impeded by utility financing problems, and by environmental restrictions on the use of coal. Due to its own financial constraints, the coal industry may be unable to expand quickly enough to satisfy any rapid increase in demand. Furthermore, given current environmental legislation, the industry may overdevelop surface resources at the expense of the underground coal industry, leading to coal shortages later as recoverable surface resources are depleted.

To avoid future constraints on the use of coal, an Accelerated Coal program (outlined in Figure 7-2) has been designed with the aid of the COAL2 model. Briefly, the program consists of: (1) financial policies designed to increase energy investments above anticipated levels by regulating oil and gas prices at higher levels, utility rate-relief, and coal price supports and (2) policies designed to channel the increased investments into coal-based energy sources such as accelerated synthetic fuels development, accelerated construction of coal-fired utilities, and accelerated development of underground coal. (A detailed description of the changes in COAL2 made to represent the Accelerated Coal program can be found in Chapters Four, Five, and Six.)

Figure 7-9 illustrates the behavior of the COAL2 model when the Accelerated Coal program is started in 1977. The model's behavior is substantially altered as coal, not imports, becomes the dominant source of energy over the next 35 years. Oil imports peak in 1985 at about 35 quads per year (17 mbpd), 35 percent of gross energy inputs. Imports drop to negligible levels by 2000, when the U.S. is dependent on coal for about two-thirds of its energy.

The behavior of each of the energy-supply sectors reflects an increased dependence on coal in this scenario. In the oil and gas sector (Figure 7-9d), demand is substantially reduced from the

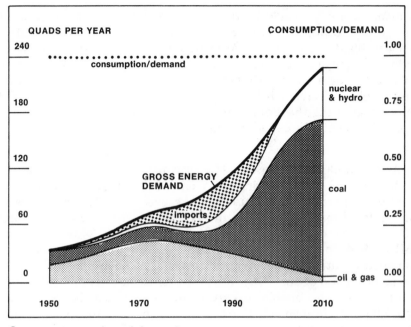

a. Gross energy supply and demand

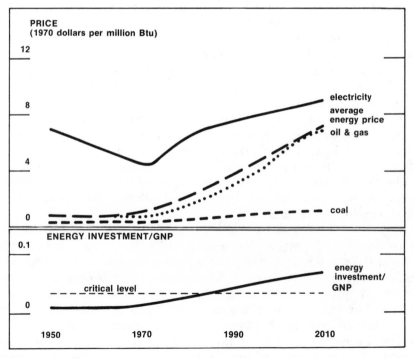

b. Economic effects

Figure 7-9.   Accelerated Coal projection

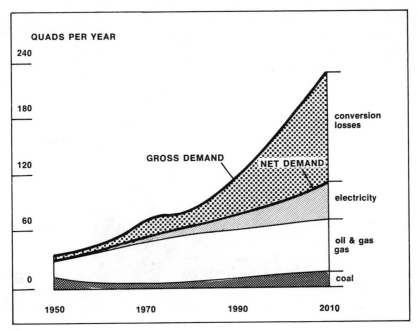

c. Demand sector

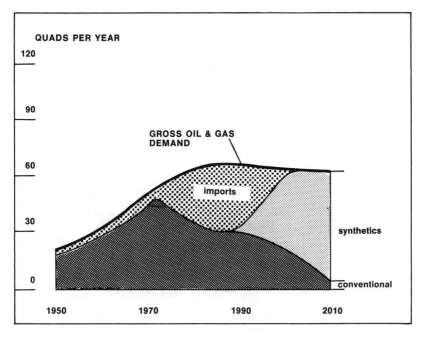

d. Oil and gas sector

Figure 7-9 (continued).   Accelerated Coal projection

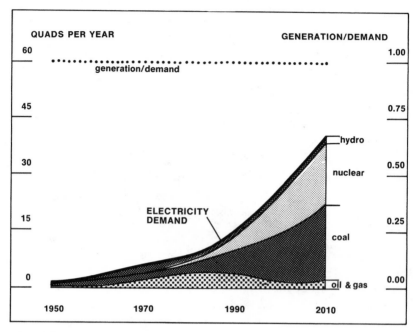

e. Electricity sector

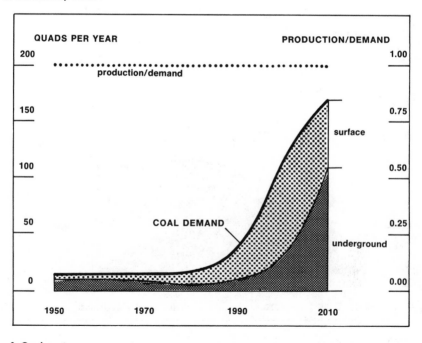

f. Coal sector

Figure 7-9 (continued).   Accelerated Coal projection

reference run due to increased prices and reduced utilities demand. Higher oil and gas prices also stimulate increased production from conventional wells, and an accelerated shift to synthetics. By 2010, 90 percent of oil and gas demand is satisfied from synthetic plants producing over 50 quads per year of oil and gas substitutes. Increases in oil and gas prices also stimulate investment in nuclear power. In Figure 7-9e, nuclear power satisfies 37 percent of electricity demands by 2000, compared to 26 percent in the reference run (and only 32 percent with the Accelerated Nuclear program!). Coal-fired utilities satisfy the bulk of remaining demand for electricity, due to the availability of SO_2 emissions control technologies (such as stack gas scrubbers). In Figure 7-9, the rapid increase in coal demand (from synthetic-conversion facilities and coal-fired utilities) is met by increases in both surface and underground mine production. The introduction of a price-support policy for coal and the restrictions imposed on surface mining allow a rapid and sustained expansion of the coal industry.

Although the Accelerated Coal program substantially improves the U.S. imports position, the U.S. economy could not, in all likelihood, accommodate the substantial increases in energy investments stimulated by the policy changes. Energy investments create a capital shortage by 1985 and grow to over 15 percent of GNP by 2010 (Figure 7-9b), an amount greater than the current 10 percent share of GNP allocated to *total* business fixed investments (FEA 1974, p. 293). Although the macroeconomic effects of this increase are not included in the COAL2 model, the implied shift in investment could be sustained only with severe depressive effects on GNP and with gross structural unemployment.

Supply Policies Summary. While the Accelerated Nuclear program leaves the reference behavior mode of the COAL2 model unchanged (imports continue to rise through the year 2000), the Accelerated Coal program is an effective long-term supply policy. When coal use is accelerated, imports peak in 1985 and drop to negligible levels by 2000. Coal is more effective as a transition fuel than nuclear power for the following reasons:

1. *Coal builds upon a broader, better established energy base.* Coal provided 18 percent of gross U.S. energy inputs in 1975, while nuclear power provided only about 2 percent (FEA 1976, p. xxvi). With the Accelerated Coal program, coal could exceed 50 percent of U.S. energy supplies by 2000, while even higher growth rates allow nuclear power to capture only 20 percent by the same year.
2. *Coal can be converted into oil and gas.* Electrical energy (and

process steam) is the only form of energy which nuclear reactors now produce.[c] Yet the energy crisis springs from a shortage of oil and gas. In the Accelerated Nuclear run, over 50 percent of final demand in 2010 still must be satisfied by oil and gas.

3. *Coal resources are more abundant.* Without rapid deployment of the breeder reactor, uranium resources are not sufficient to support the Accelerated Nuclear Program (FEA-NTF 1974, p. 2.0-6; Day 1975, p. 52). Recoverable coal resources represent 30 times the Btu content of recoverable uranium resources (*Coal Facts* 1974, p. 7).

4. *The environmental side effects of coal are more amenable to technological solutions.* The major environmental side effects associated with coal production and use—strip mine damage, underground miner health and safety, and SO_2 emissions—can be reduced through the use of proper, although more expensive, technologies. Nuclear environmental consequences—accident risks, danger of sabotage, and difficulties of radioactive waste disposal— currently seem highly resistant to technological solutions (Weinberg 1972; Kneese 1973).

Although the Accelerated Coal program substantially improves the long-term balance between U.S. energy supply and demand, severe, persistent economic dislocations would result without a parallel program to reduce demand. Macroeconomic effects are not included explicitly in the COAL2 model, but it is likely that the economic risks of such a program would be as great as the political risks of no program, as represented by the reference projection's maximum dependence on imports.

Combined Supply and Demand Policies

The COAL2 policy simulations presented in the previous two sections support the conclusion that neither demand nor supply policy options *alone* can engender a satisfactory U.S. energy transition.

The two following simulations combine the Accelerated Coal program with the Accelerated Conservation and Zero Energy Growth programs. The energy demand policies are designed to reduce the need for domestic energy investment to the point where the Accelerated Coal program's undesirable economic side effects could be tolerated.

---

[c]Some studies suggest the creation of hydrogen from off-peak nuclear operations. Yet the large-scale use of such processes is not expected to be feasible until after 2000, too late to alleviate the worsening oil and gas shortage of the transition period (Manne 1975, p.12).

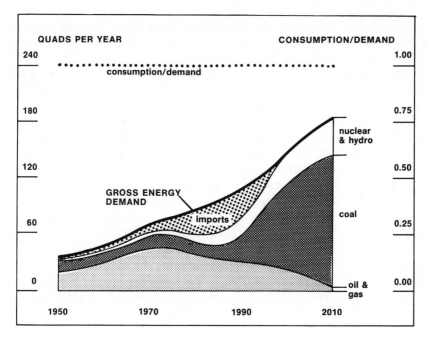

a. Gross energy supply and demand

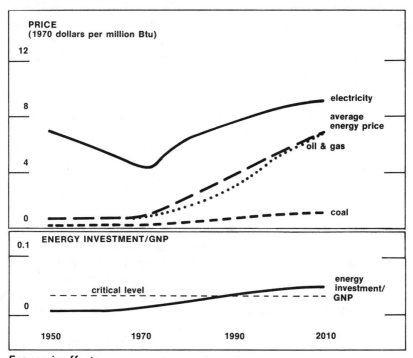

b. Economic effects

Figure 7-10. Combined Coal and Conservation projection

Combined Coal and Conservation Projection. If the federal government formulates a national energy program, the program objectives are likely to be designed to promote technological solutions to the energy problem. Figure 7-10 shows the behavior of the COAL2 model when the most effective technological policies affecting both supply and demand are combined.

The Combined Coal and Conservation programs together reduce net energy demand about 10 percent below demand in the Accelerated Conservation policy run alone (Figure 7-5). These additional energy savings are induced by higher energy prices. Some of these savings are lost when gross demands are compared, for the Accelerated Coal program results in a greater use of synthetic fuels and electricity, both of which involve large conversion losses. The addition of the Accelerated Coal policies produces a substantial long-term improvement in model behavior. Imports are reduced essentially to zero by 2000 with the combined technological policies, while in both the reference run and the Accelerated Conservation projection imports still satisfy close to 40 percent of gross energy demand by 2000.

Unfortunately, Figure 7-10 indicates that the U.S. economy will probably not be able to afford a technological solution. The investments required to satisfy a "technically-fixed" energy demand will severely strain the capital market. Heavy investment requirements for electric utilities and synthetic fuel facilities increase energy's share of the GNP budget beyond the maximum tolerable level (7 percent) after 1985.

A more basic problem associated with a strict technological approach to the solution of the U.S. energy imbalance stems from the tendency for technological solutions simply to postpone the recurrence of problems similar to the current oil and gas crisis. Such is the case here: with the technological solutions tested in Figure 7-10, U.S. energy consumption is still growing in the year 2010, and is still dependent on a finite energy resource, coal. These are the very characteristics causing the current crisis.

Combined Zero Energy Growth and Accelerated Coal Projection. In Figure 7-11, the technological policies of Figure 7-10 (Combined Coal and Conservation programs) are combined with policies that tend to stabilize growth in energy demand over the long term. In COAL2, this policy is represented by gradually stabilizing GNP, the measure of material consumption in the model, by 2010 (as in Figure 7-6).

Due to the combined effects of demand reductions from price increases, Accelerated Conservation policies, and Zero Energy

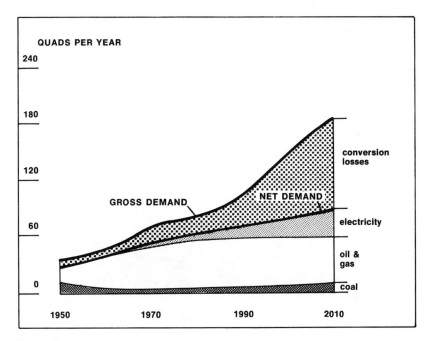

c. Demand sector

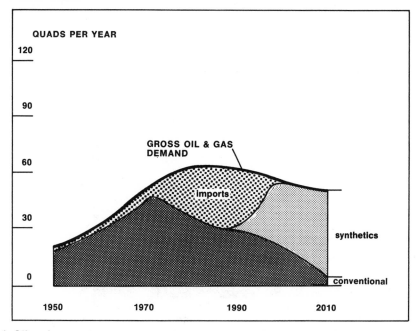

d. Oil and gas sector

Figure 7-10 (continued). Combined Coal and Conservation projection

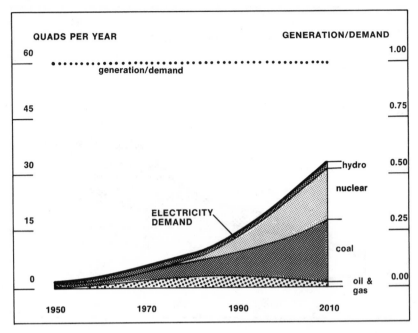

e. Electricity sector

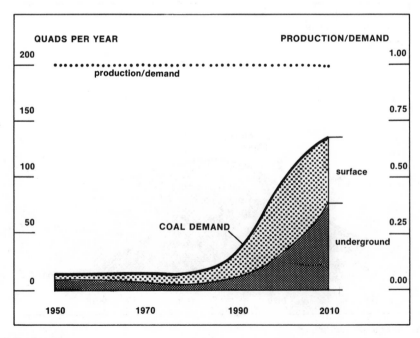

f. Coal sector

Figure 7-10 (continued).   Combined Coal and Conservation projection

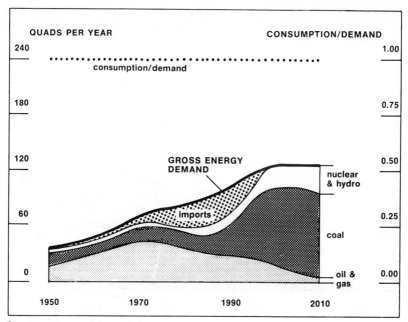

a. Gross energy supply and demand

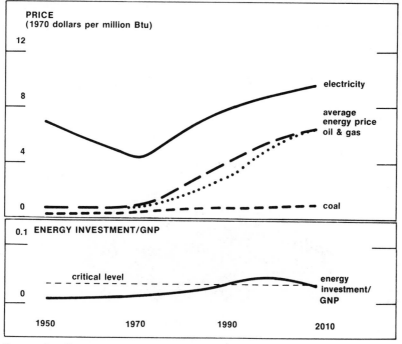

b. Economic effects

Figure 7-11. Combined Zero Energy Growth and Accelerated Coal projection

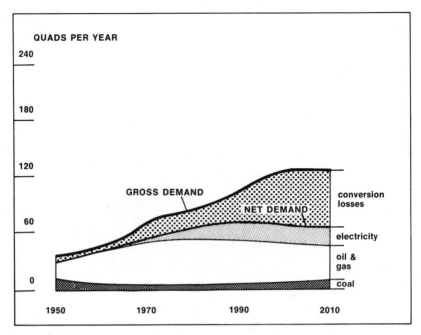

c. Demand sector

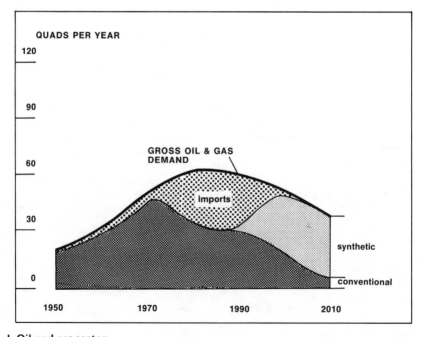

d. Oil and gas sector

Figure 7-11 (continued).   Combined Zero Energy Growth and Accelerated Coal
projection

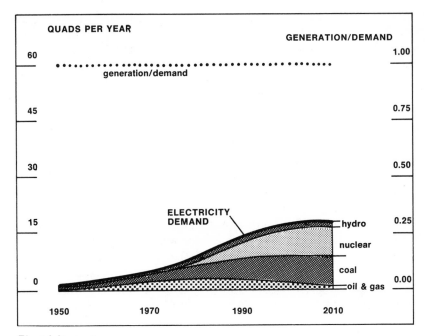

e. Electricity sector

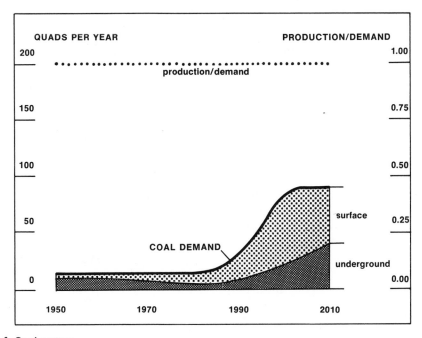

f. Coal sector

Figure 7-11 (continued).   Combined Zero Energy Growth and Accelerated Coal
projection

Growth policies, gross energy demand stabilizes in 2000 at 125 quads per year. In Figure 7-11d, demand for oil and gas stabilizes near current levels for 15 years after 1975. Oil and gas demand begins to drop after 1990, allowing oil imports to peak in 1985 and drop sharply thereafter. Because ZEG does not significantly affect growth in energy demand until after 1990, 1985 dependency on imports is essentially the same as the previous run. In both cases, 35 percent of gross energy inputs (and 50 percent of U.S. oil and gas supplies) must still be imported in 1985.

Gradual stabilization of U.S. energy demand tends to remove the two major problems that plagued the Combined Coal and Conservation projection: (1) Energy investments average only about 6 percent of GNP during the transition period, approximately double their current share. A 1974 Ford Energy Policy Project substudy states that such an increase could probably be absorbed by the U.S. capital markets (Hass, Mitchell, and Stone 1974, p. 106). (2) Coal production, stabilized at 90 quads per year (4 billion tons/year) in 2000, could be maintained for another 25 to 50 years beyond the year 2000 with no major cost increases. The extra time would allow the United States to shift gradually to ultimate, nondepletable energy sources such as solar or fusion power.

COMBINED POLICIES SUMMARY

When energy supply policies—the Accelerated Coal policies outlined in Figure 7-2—are combined with policies that reduce demand below "business-as-usual" levels—Accelerated Conservation and Zero Energy Growth—the behavior of the U.S. energy system is substantially improved. Imports peak in 1985 and decline thereafter. By 2000 the nation achieves energy independence. However, reliance on a purely technical solution to the U.S. energy problem (Figure 7-10) requires energy investments too massive to be absorbed by U.S. capital markets. The addition of a program that stabilizes U.S. energy demand (Figure 7-11) creates an energy transition that is both economically affordable and consistent with a long-term goal of a smooth transition to ultimate, nondepletable energy sources.

Even under the most optimistic energy policy scenario, the COAL2 model projects substantial increases in oil and gas imports through 1985. Both of the combined supply and demand policy runs (Figures 7-10 and 7-11) project an increase in oil imports to over 30 quads per year in 1985, 2.5 times current levels. Dependence on imports over the short term is caused by the long time lags in implementing any major policy change in the energy supply or demand systems. The COAL2 model shows that even with major energy policy changes, 10 years are required to reverse the upward trend in imports. Consequently, the "Project Independence" goals set by the FEA, ERDA, and other energy studies (FEA 1976, p. xxvii; ERDA 1976, p. vii; Teller 1975, p. 11) are unachievable. Changes in energy policy, although badly needed, can have little effect on the behavior of the U.S. energy system over the next 10 years.

✳ *Chapter 8*

The Policy Implications
of COAL2

The rapid rise in United States dependence on foreign oil imports and the concurrent oil embargo and quadrupling of OPEC oil prices in 1973 have created the high-priority national goal of Project Independence. This phrase has been applied to any set of energy strategies designed to reduce imports over the long term. Although current energy analyses disagree on which policies to emphasize, the general consensus of the Federal Energy Administration, the Energy Research and Development Administration, and most other research groups is that oil and gas imports could be reduced below current levels by 1985 (FEA 1976, p. xxvii; ERDA 1975, p. S-5; Ford 1974a, p. 76; MIT 1974; Teller 1975, p. 15). In contrast, using a dynamic simulation model of energy supply and demand (COAL2), this study concludes that the United States energy problem is fundamentally more severe than other analyses and agencies indicate, and much less amenable to a near-term solution (Figure 8-1).

With no major redirections in domestic energy policy, the United States energy system will soon exhibit a major imbalance between domestic energy supply and demand (Figure 8-2). Depletion of recoverable United States oil and gas resources causes domestic production to peak and decline after 1972. Because of the high cost of alternatives, research and development delays, and environmental considerations, there are few incentives for investing in other domestic fuel sources, and hence gross domestic energy production can barely be kept near current (1975) levels. As energy demand rises, a growing fraction of total energy inputs must be imported. By 1985 oil imports reach 36 quads per year, three times current levels.

- Given no major changes in energy policy (the **reference projection**), the United States must import almost 40% of its energy from 1985 to 2000. Figure 7-3.

- If **energy demand stabilized** by the year 2000, imports would peak in 1985 at 33 quads per year (35% of consumption), and drop to zero by 2010. Figure 7-6.

- An **Accelerated Nuclear program** is ineffective in reducing United States dependence on imports; with it, imports are reduced only 10% below the reference projection by the year 2000. Figure 7-8.

- An **Accelerated Coal program** could achieve the national goal of independence from foreign oil imports, but not until the year 2000. Yet this program could cause severe economic dislocations due to the major shifts in investment to the energy sector. Figure 7-9.

- A **combined program** that stabilizes energy demand over the long term and accelerates the use of coal (the Zero Energy Growth and Accelerated Coal programs) generates a smooth transition that balances United States energy supply and demand by the year 2000. Figure 7-11.

- Under **any policy circumstances**, United States dependence on imports increases to 1985, energy prices increase substantially above 1975 levels, and much investment—and foresight—will be required of citizens, government, and industry during the next 35 years.

Figure 8-1. Major conclusions of the COAL2 study

Energy imports supply almost 50 percent of gross United States energy consumption in 1985, compared to current (1975) dependence of about 20 percent. In the reference projection of COAL2, import dependency *worsens* after 1985 to a total of 60 quads per year in 2000. It is unlikely that the strategic vulnerability and balance of payments problems caused by such massive dependence on foreign oil could be tolerated.

Four major policy programs (two demand programs: Accelerated Conservation and Zero Energy Growth, and two supply programs: Accelerated Nuclear and Accelerated Coal) have been tested on the COAL2 model. The effects of these policy tests on imports in two future test years, 1985 and 2000, are shown in Figures 8-3 and 8-4. The policy analysis indicates that even if energy demand is stabilized, energy independence could not be reached by the year 2000. Because of the rapid decline in oil and gas production (a 40 percent reduction from the 1972 peak by 1985), it is too late to solve the transition problem simply by reducing demand. Even to maintain current consumption levels, alternative domestic energy supplies must be developed at an accelerated pace.

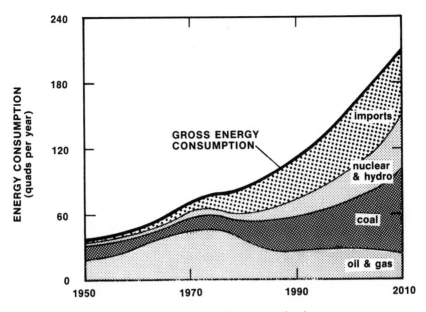

Figure 8-2. COAL2 reference projection

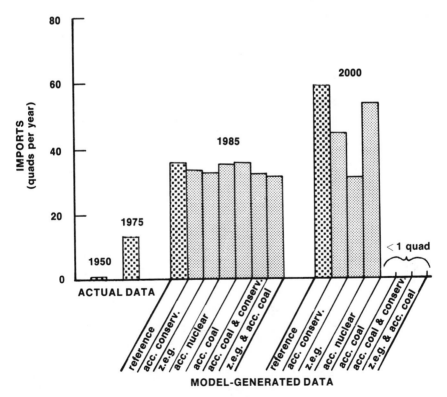

Figure 8-3. Effects of policy tests on imports

Policy	Figure Number	1985					2000				
		Gross Energy Consumption (quads)	Percent Satisfied by				Gross Energy Consumption (quads)	Percent Satisfied by			
			Imports	Conv. Oil & Gas	Coal	Nuclear & Hydro		Imports	Conv. Oil & Gas	Coal	Nuclear & Hydro
Reference	7-3	99	37%	29%	24%	10%	155	38%	18%	27%	17%
Demand											
1) Accelerated Conservation	7-5	94	35	31	24	10	136	33	21	29	17
2) Zero Energy Growth	7-6	94	35	31	24	10	115	27	24	31	18
Supply											
3) Accelerated Nuclear	7-8	99	35	29	23	13	155	35	18	27	20
4) Accelerated Coal	7-9	98	36	31	22	11	174	—	13	67	20
Combined											
5) Accelerated Coal & Conservation	7-10	93	35	32	22	11	149	—	13	67	20
6) Zero Energy Growth & Accelerated Coal	7-11	92	34	33	22	11	127	—	14	66	20

Figure 8-4. Tabular summary of policy tests

A program designed to reduce imports by accelerating the development of nuclear power (the Accelerated Nuclear Program) proves ineffective. Even though nuclear capacity were to expand at a rate close to the maximum FEA projection, nuclear power would provide only 20 percent of gross energy inputs by the year 2000. Furthermore, an Accelerated Nuclear program only tends to substitute nuclear power for coal during the transition period, because these two fuels dominate new utility construction after 1973. Imports of oil and gas are hardly affected.

Only the Accelerated Coal Program effectively reduces U.S. dependence on imports after 1985. Yet the COAL2 model demonstrates that an attempt to maintain high growth rates in energy consumption with accelerated supply programs could create severe economic disruption (see Figure 7-9b).

United States energy policymakers are the first to admit that there is no easy solution to the U.S. energy problem. The analysis outlined in this book suggests that *the difficulties in establishing an effective U.S. energy policy are an inherent property of the energy system,* not simply a case of politicians unable to make up their minds. It is a general property of complex systems that there is often a fundamental conflict between the short-term and long-term consequences of any policy change (Forrester 1971, p. 7). In the U.S. energy system, *any* energy policy action creates a controversy between short-term and long-term interests.

For example, the COAL2 model indicates that the policy changes that improve the long-term energy picture are likely to have negative short-term side effects. Energy prices, capital investment, and environmental damage all increase substantially before 1990 when Zero Energy Growth and Accelerated Coal policies replace the trend projections of the reference run. Such is the price the system exacts for the long-term benefits of reduced imports that accrue after 1990.

Conversely, policies designed to improve the behavior of the energy system over the short run will often worsen our energy problems over the long run. The reference run of the COAL2 model, representing current U.S. energy policy, reflects this property. In an effort to continue to supply U.S. consumers with low-cost energy in the short run, current energy policy will almost certainly lead to a post-1985 period of uncertain supplies, excessively high energy prices, and political and economic disruption due to massive dependence on imports.

It is an unfortunate fact of the current political decision-making process that short-term considerations most often dominate policy considerations. For example, the oil price rollback called for in the

Energy Policy and Conservation Act of 1975 was designed to avoid the short-term shock to the U.S. economy of a sudden increase in energy prices due to decontrol of old oil. The short run is always more visible and compelling. It speaks loudly for immediate action. Yet a continuation of current policies aimed at short-term improvement could lead to long-run economic and political consequences—severe recession, energy shortages, or a war for the control of Arab oil—that are totally unmanageable and unacceptable.

We conclude that the nation's long-term interests are best served by reducing energy demand and accelerating the use of coal, in spite of the short-run economic and environmental sacrifices that accompany these policies. Even if these policies are not carried out exactly as suggested, we hope that our modeling effort has clarified the long-term U.S. energy future and dispelled some common misunderstandings about the U.S. energy problem. If the model can achieve this, long-term considerations will weigh more heavily in the decision making process—as they properly should.

Appendix A
Establishing Confidence
in COAL2

To the policymaker, a model is a representation of reality. The model is useful to the extent that it behaves, *for his purpose*, like the real world. Modeling authorities have repeatedly affirmed the impossibility of any absolute criteria for establishing the "validity" of a model. Instead, what is possible (and necessary) is to understand a model's utility:

> When a model is taken as a substitute for reality there can be no objective proof of its validity. *Validity means not absolute truth, but only a degree of confidence.* The operator's (policymaker's) model of a social system is like the engineer's theory of heat transfer or his theories about the strength of materials. These physical theories (including physical science theories such as Einstein's law) rest on no foundation that permits an absolute proof. They rest only on a foundation of confidence that has been generated by repeated demonstrations that the theories serve a useful purpose and have not been shown to be invalid for the purposes to which they are expected to apply. Validity in models is a subjective matter that is always judged by estimating how much evidence is necessary in a particular case to establish sufficient confidence to justify taking action on the model.
>
> Forrester 1973, p. 27

A model's validity can be judged only in the context of its purpose. No model is good or bad in an absolute sense; it is either appropriate or inappropriate in the context of some use (for example, see Naill 1974, p. 65). In an examination of the validity of the COAL2 model, energy policymakers must be convinced that the

209

reference behavior shown in Figure 7-3 is credible, and, more important, that policy changes would have real-world effects similar to those indicated by the model. To increase confidence in the COAL2 model, it has been subjected to three common validity tests (Forrester 1973, p. 48).

STRUCTURE VERIFICATION TEST

To verify the structure of a model, that structure must be compared closely with the real system's structure to establish whether the model does, in fact, represent the essential features of reality which are of interest to the policymaker. With models as large as COAL2— which has 12 explicit state variables—this detailed comparison is usually not feasible. In the case of COAL2, however, we have taken great care to make the presentation in Chapters Three through Six in such a way that even those unfamiliar with modeling can expect to judge the quality of the model. Using Volume II of this book, those with mathematical or modeling backgrounds will be able to subject COAL2 parameters to similar comparisons with real-world data.

MISTAKEN IDENTITY TEST

A dynamic model of the United States energy system should behave like the system it represents. The model-generated time series data should have the same behavioral characteristics as the time series data from the real system. The mistaken identity test asks whether a person familiar in detail with the real system might mistake model-generated data for that of the real system.

Figures A-1 and A-2 compare real-world historical data of the United States energy system with behavior of the COAL2 model over the same time period. Bear in mind that these historical time-series data were *not* used to construct the model. A comparison of Figures A-1 (historical data) and A-2 (COAL2 data) confirms that the COAL2 model generates time-dependent behavior very close to the dynamic characteristics of the real United States energy system.

PARAMETER SENSITIVITY TEST

The reference projection of the COAL2 model (Figure 7-3) exhibits a marked propensity toward increased imports after 1970 due to depletion of domestic oil and gas resources and the lack of sufficient incentives for the development of an alternative domestic energy source. To what degree is this projection disturbed by uncertainties in the parameter values in the model? If the reference behavior and

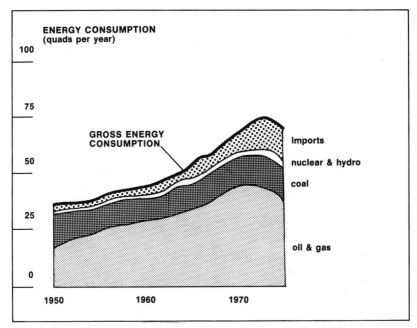

a. Gross energy supply and demand

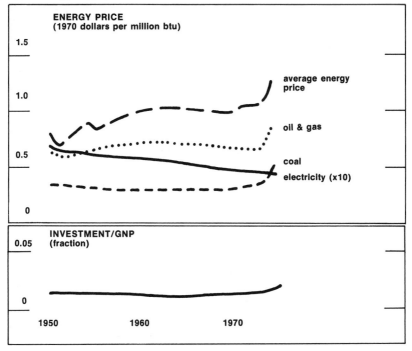

b. Economic effects

Compare with the corresponding plot of Figure A-2. Can you tell which is model-generated data and which is historical data? See text (Mistaken Identity Test) for answer.

Figure A-1.   COAL2 output or historical data?

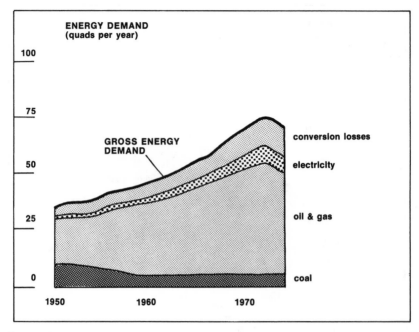

c. Demand sector

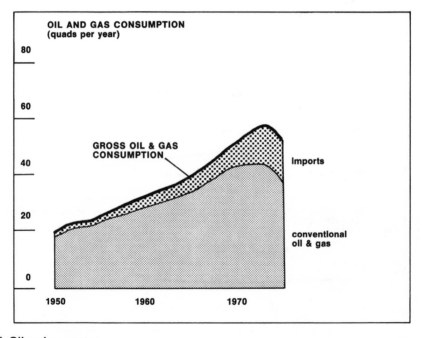

d. Oil and gas sector

Figure A-1 (continued). COAL2 output or historical data?

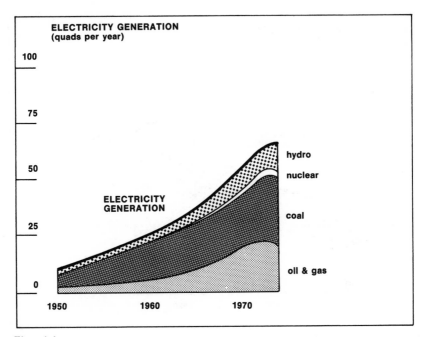

e. Electricity sector

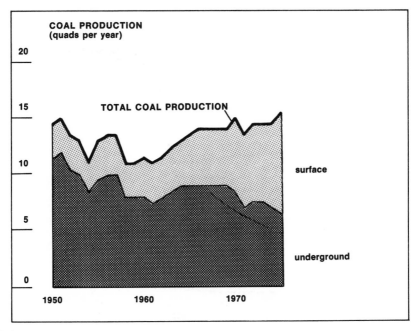

f. Coal sector

Figure A-1 (continued).   COAL2 output or historical data?

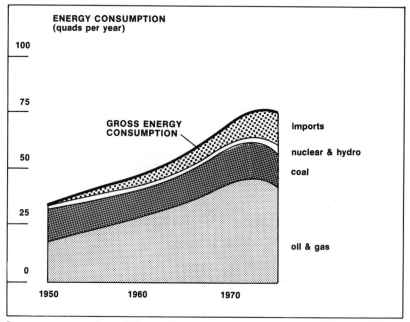

a. Gross energy supply and demand

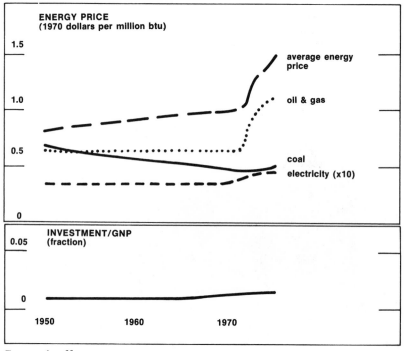

b. Economic effects

Compare with the corresponding plot of Figure A-1. Can you tell which is model-generated data and which is historical data? See text (Mistaken Identity Test) for answer.

Figure A-2. COAL2 output or historical data?

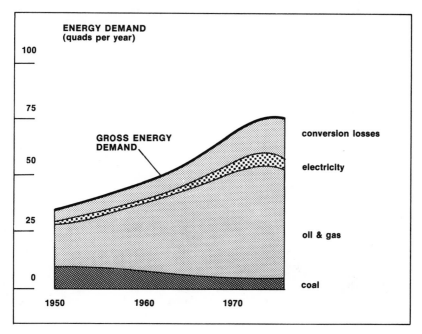

c. Demand sector

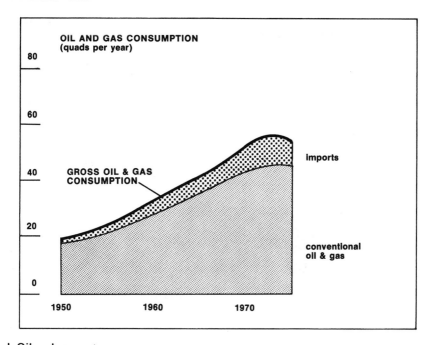

d. Oil and gas sector

Figure A-2 (continued).   COAL2 output or historical data?

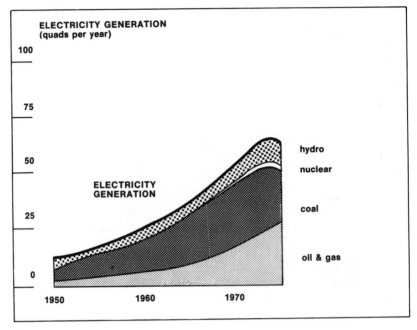

e. Electricity sector

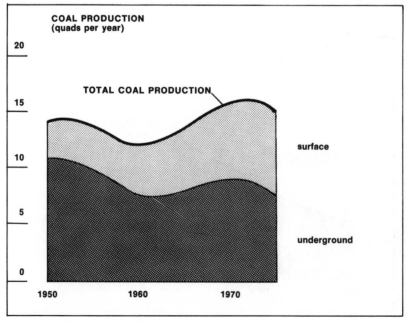

f. Coal sector

Figure A-2 (continued). COAL2 output or historical data?

policy tests of Chapter Seven were relatively insensitive to parameter uncertainty, the risk involved in adopting the model for policy-making would be substantially reduced.

Figures A-3 and A-4 show the results of sensitivity tests on critical parameters in the COAL2 model. Eleven parameters[a] were varied by a factor of two (high or low), establishing a rough, widely-defined confidence interval around the COAL2 projections.

In lieu of reproducing plotted output for each of the sensitivity tests, Figures A-3 and A-4 focus on the sensitivity of COAL2's oil input projections in two test years, 1985 and 2000. As might be expected, the model's predictions of import levels become more sensitive to parameter uncertainties as time progresses (Figure A-3). Yet even with uncertainties, the model's policy conclusions remain

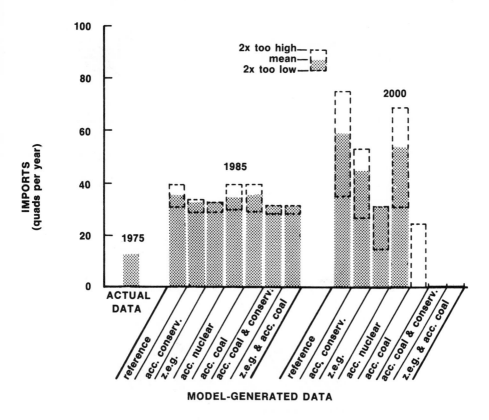

Figure A-3. Import sensitivity, in quads per year

---

[a]GNP growth rate, total demand elasticity, oil and gas resources, synthetic fuel costs, synthetic investment decision, nuclear capital costs, nuclear fuel costs, fossil-fired utilities capital costs, coal pricing decision, underground labor hiring delay, surface coal resources.

intact. Imports are never reduced below 28 quads per year in 1985 under any circumstances. Figure A-4 shows that the relative effectiveness of each policy option is unchanged by uncertainty. The insensitivity of the COAL2 model behavior to parameter uncertainties seems surprising at first. Yet, upon closer examination, none of these parameter changes affect the basic structural elements responsible for the model behavior: rapid depletion of the finite stock of domestic oil and gas resources coupled with delays and restrictions on the incentives to develop alternative domestic energy sources. As shown in Figures A-3 and A-4, substantial changes in United States energy policy are needed to alter the behavior projected by the COAL2 model.

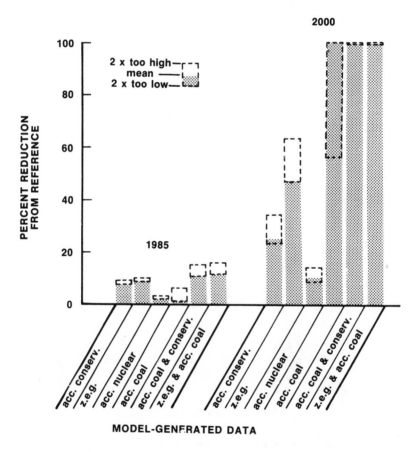

MODEL-GENERATED DATA

Figure A-4.   Import sensitivity, in percentage change from reference run

Appendix B
COAL2 Program Listing

```
100 *      COAL2: DARTMOUTH ENERGY POLICY MODEL
110 NOTE
150 NOTE
160 NOTE   TOTAL ENERGY DEMAND
170 NOTE
180 A      NED.K=EGNPR70*GNP.K*DMP.K
190 C      EGNPR70=5.77E4
200 L      GNP.K=GNP.J+(DT)(GNPIR.JK)
210 N      GNP=GNPI
220 C      GNPI=4.81E11
230 R      GNPIR.KL=GNP.K*GNPGR.K
240 A      GNPGR.K=CLIP(GNPGR1.K,LTGR.K,TIME.K,RSYEAR)
250 C      RSYEAR=1973
260 A      GNPGR1.K=CLIP(LTGR.K,RYGR.K,TIME.K,RCYEAR)
270 C      RCYEAR=1975
280 A      LTGR.K=TABLE(LTGRT,TIME.K,1950,2010,10)*1E-2
290 T      LTGRT=3.55/3.55/3.55/3.4/3.2/3/3/2.8
300 A      RYGR.K=TABHL(RYGRT,TIME.K,1974,1976,1)*1E-2
310 T      RYGRT=-2.1/-3.6/3.5
320 A      DMP.K=SMOOTH(IDMP.K,DAT)
330 N      DMP=1.09
340 C      DAT=10
350 A      IDMP.K=CLIP(IDMP2.K,IDMP1.K,TIME.K,PYEAR)
360 A      IDMP1.K=TABHL(IDMP1T,AEP.K/AEPN,0,10,1)
370 T      IDMP1T=1.2/1/.82/.74/.68/.64/.61/.58/.56/.54/.52
380 A      IDMP2.K=TABHL(IDMP2T,AEP.K/AEPN,0,10,1)
390 T      IDMP2T=1.2/1/.82/.74/.68/.64/.61/.58/.56/.54/.52
400 C      AEPN=.94E-6
410 A      AEP.K=(AOGP.K*NOGD.K*OGCDR.K+EP.K*TEG.K+
420 X      CPRICE.K*DCUD.K*CPDR.K)/NEC.K
430 A      NEC.K=NOGD.K*OGCDR.K+TEG.K+DCUD.K*CPDR.K
440 NOTE
450 NOTE   INTERFUEL SUBSTITUTION
460 NOTE
470 A      FEDC.K=TABHL(FEDCT,GNP.K/GNP70,.5,1.9,.2)*CDSM.K
480 T      FEDCT=.35/.15/.105/.087/.07/.06/.055/.05
490 C      GNP70=974E9
500 A      CDSM.K=TABLE(CDSMT,SCOPR.K/SCOPR70,0,2,.2)
```

219

```
510 T      CDSMT=5/4.5/2.5/1.7/1.3/1/.83/.71/.63/.56/.5
520 C      SCOPR70=.52
530 A      SCOPR.K=SMOOTH(COPR.K,DAT)
540 N      SCOPR=.54
550 A      COPR.K=CPRICE.K/AOGP.K
560 A      DCUD.K=FEDC.K*NED.K
570 A      FEDE.K=TABHL(FEDET,GNP.K/GNP70,0,8,1)*EDSM.K
580 T      FEDET=.03/.093/.14/.18/.21/.24/.26/.275/.28
590 A      EDSM.K=TABLE(EDSMT,SEPR.K/SEPR70,0,2.5,.25)
600 T      EDSMT=2.5/1.9/1.5/1.22/1/.78/.64/.56/.5/.44/.4
610 C      SEPR70=8.75
620 A      SEPR.K=SMOOTH(EPR.K,DAT)
630 N      SEPR=13.5
640 A      EPR.K=EP.K/AOGP.K
650 A      NELD.K=FEDE.K*NED.K
660 A      NOGD.K=(1-FEDE.K-FEDC.K)*NED.K
670 NOTE
680 NOTE   *** OIL AND GAS SECTOR ***
690 NOTE
700 NOTE   OIL AND GAS SUPPLY-DEMAND BALANCE
710 NOTE
720 A      GOGD.K=NOGD.K+OGDU.K
730 A      DOGCUF.K=TABHL(DOGCUFT,GOGD.K/DOGPC.K,0,1.6,.2)
740 T      DOGCUFT=0/.2/.4/.6/.8/.85/.87/.89/.9
750 A      DOGPC.K=COGPC.K+SOGPC.K
760 A      OGCDR.K=OGC.K/GOGD.K
770 A      OGC.K=DOGPR.K+OGI.K
780 A      DOGPR.K=COGPR.K+SOGPR.K
790 NOTE
800 NOTE   OIL AND GAS FINANCING
810 NOTE
820 A      AOGP.K=(DOGP.K*DOGPR.K+OGIP.K*OGI.K)/OGC.K
830 A      DOGP.K=CLIP(UOGP.K,ROGP.K,TIME.K,DEREGT)
840 C      DEREGT=4000
850 A      ROGP.K=RPRAT.K*DCOST.K
860 A      RPRAT.K=CLIP(PPRAT,HPRAT,TIME.K,PYEAR)
870 C      HPRAT=1.86
880 C      PPRAT=1.86
890 A      UOGP.K=MIN(UPRAT.K*DCOST.K,OGIP.K)
900 A      UPRAT.K=TABHL(UPRATT,DOGPC.K/GOGD.K,0,2,.25)
910 T      UPRATT=6/5.4/4/3/2.4/1.86/1.5/1.2/1
920 A      DCOST.K=(CCOST.K*COGPR.K+SCOST.K*SOGPR.K)/DOGPR.K
930 A      DOGREV.K=DOGP.K*DOGPR.K
940 A      POGRR.K=CLIP(POGRR2.K,POGRR1.K,TIME.K,PYEAR)
950 A      POGRR1.K=TABHL(POGRR1T,AOGROI.K,0,.20,.02)
960 T      POGRR1T=0/.02/.05/.08/.13/.22/.37/.52/.57/.59/.6
970 A      POGRR2.K=TABHL(POGRR2T,AOGROI.K,0,.20,.02)
980 T      POGRR2T=0/.02/.05/.08/.13/.22/.37/.52/.57/.59/.6
990 A      AOGROI.K=SMOOTH(DOGROI.K,ROIAT)
1000 C      ROIAT=5
1010 A      DOGROI.K=(DOGP.K-DCOST.K)*DOGPR.K/(COGC.K+SOGC.K)
1020 A      DOGINV.K=POGRR.K*DOGREV.K
1030 NOTE
1040 NOTE   CONVENTIONAL OIL AND GAS PRODUCTION
1050 NOTE
1060 L      COGC.K=COGC.J+(DT)(COGICR.JK-COGDR.JK)
1070 N      COGC=COGCI
1080 C      COGCI=45.8E9
1090 R      COGDR.KL=COGC.K/ALOGC
1100 C      ALOGC=20
1110 R      COGICR.KL=DELAY3(COGIR.JK,COGCT)
1120 C      COGCT=5
```

```
1130 R      COGIR.KL=DOGINV.K*(1-FIASS.K)
1140 A      CCOST.K=OGCCAF*(1/COGCCR.K)
1150 C      OGCCAF=.175
1160 A      COGCCR.K=CCCRI*OGCEM.K
1170 C      CCCRI=.5E6
1180 A      OGCEM.K=TABLE(OGCEMT,FOGRR.K,0,1,.1)
1190 T      OGCEMT=0/.04/.09/.16/.26/.6/1/1/1/1/1
1200 A      FOGRR.K=COGR.K/COGRI
1210 C      COGRI=2.8E18
1220 L      COGR.K=COGR.J+(DT)(-COGDPL.JK)
1230 N      COGR=COGRI-COGP50
1240 C      COGP50=.46E18
1250 R      COGDPL.KL=COGPR.K
1260 A      COGPR.K=COGPC.K*DOGCUF.K
1270 A      COGPC.K=COGCCR.K*COGC.K
1280 NOTE
1290 NOTE   SYNTHETIC OIL AND GAS PRODUCTION
1300 NOTE
1310 L      SOGC.K=SOGC.J+(DT)(SOGICR.JK-SOGCDR.JK)
1320 N      SOGC=SOGCI
1330 C      SOGCI=0
1340 R      SOGCDR.KL=SOGC.K/ALSYNC
1350 C      ALSYNC=20
1360 R      SOGICR.KL=DELAY3(SOGIR.JK,SCT)
1370 C      SCT=5
1380 R      SOGIR.KL=DOGINV.K*FIASS.K
1390 A      FIASS.K=DLINF3(IFIASS.K,SDT)
1400 C      SDT=8
1410 A      IFIASS.K=CLIP(FIASS2.K,FIASS1.K,TIME.K,PYEAR)
1420 A      FIASS1.K=TABHL(FIASS1T,CCOST.K/SCOST.K,0,2,.25)
1430 T      FIASS1T=0/0/0/.1/.5/.8/.9/.95/1
1440 A      FIASS2.K=TABHL(FIASS2T,CCOST.K/SCOST.K,0,2,.25)
1450 T      FIASS2T=0/0/0/.1/.5/.8/.9/.95/1
1460 A      SCOST.K=SPC.K+SFC.K
1470 A      SPC.K=OGCCAF*(1/SYNCCR)
1480 C      SYNCCR=.14E6
1490 A      SFC.K=CPRICE.K/SCE
1500 C      SCE=.6
1510 A      SOGPC.K=SOGC.K*SYNCCR
1520 A      SOGPR.K=SOGPC.K*DOGCUF.K*CPDR.K
1530 A      CDS.K=SOGPC.K*DOGCUF.K/SCE
1540 NOTE
1550 NOTE   OIL AND GAS IMPORTS
1560 NOTE
1570 A      OGI.K=CLIP(ROGI.K,UOGI.K,TIME.K,PYEAR)
1580 A      UOGI.K=MAX(OGD.K-DUGPCR.K,0)
1590 A      ROGI.K=MIN(OIQ*GOGD.K,UOGI.K)
1600 C      OIQ=1
1610 A      OGIP.K=CLIP(FOP2.K,FOPH,TIME.K,EMYEAR)
1620 C      FOPH=.65E-6
1630 C      EMYEAR=1973
1640 A      FOP2.K=CLIP(FOPT,FOPE,TIME.K,TYEAR)
1650 C      FOPE=2E-6
1660 C      FOPT=4E-6
1670 C      TYEAR=4000
1680 NOTE
1690 NOTE   *** ELECTRICITY SECTOR ***
1700 NOTE
1710 NOTE   ELECTRICITY SUPPLY-DEMAND BALANCE
1720 NOTE
1730 A      SED.K=NELD.K-HG.K
1740 A      SEG.K=NEG.K+CEG.K+OGEG.K
1750 A      TEG.K=SEG.K+HG.K
```

```
1760  A      SEGC.K=NEGC.K+CEGC.K+OGEGC.K
1770  A      SECF.K=TABHL(SECFT,SED.K/SEGC.K,0,1,.2)
1780  T      SECFT=0/.2/.4/.6/.77/.85
1790  A      EGDR.K=TEG.K/NELD.K
1800  NOTE
1810  NOTE   ELECTRIC UTILITY FINANCING
1820  NOTE
1830  A      EP.K=CLIP(ALELP.K,RELP.K,TIME.K,RRT)
1840  C      RRT=4000
1850  A      RELP.K=DLINF1(ALELP.K,RLT)
1860  N      RELP=7.16E-6
1870  C      RLT=2
1880  A      ALELP.K=AECOST.K+AROR.K*SEC.K/SEG.K
1890  A      AROR.K=CLIP(IAROR,NAROR,TIME.K,RRT)
1900  C      NAROR=.08
1910  C      IAROR=.10
1920  A      SEC.K=NUC.K+CUC.K+OGUC.K
1930  A      AECOST.K=(NELC.K*NEG.K+CEC.K*CEG.K+OGEC.K*OGEG.K)/SEG.K
1940  A      SER.K=EP.K*SEG.K
1950  A      FSERI.K=TABHL(FSERIT,SECF.K/SECFN.K,.7,1.1,.1)*EROIM.K
1960  T      FSERIT=0/.1/.3/.55/.7
1970  A      SECFN.K=CLIP(SECF2,SECF1,TIME.K,PYEAR)
1980  C      SECF1=.55
1990  C      SECF2=.55
2000  A      EROIM.K=TABHL(EROIMT,AEROI.K,0,.10,.02)
2010  T      EROIMT=.25/.3/.47/.85/1/1
2020  A      AEROI.K=SMOOTH(EROI.K,EROIAT)
2030  C      EROIAT=5
2040  A      EROI.K=(EP.K-AECOST.K)*SEG.K/SEC.K
2050  A      SEUI.K=FSERI.K*SER.K
2060  NOTE
2070  NOTE   HYDROPOWER GENERATION
2080  NOTE
2090  A      HG.K=TABHL(HGT,TIME.K,1950,2010,10)*1E15
2100  T      HGT=.32/.50/.84/1.1/1.3/1.38/1.4
2110  NOTE
2120  NOTE   NUCLEAR ELECTRICITY GENERATION
2130  NOTE
2140  L      NUC.K=NUC.J+(DT)(NUICR.JK-NCDR.JK)
2150  N      NUC=NUCI
2160  C      NUCI=0
2170  R      NCDR.KL=NUC.K/ALNU
2180  C      ALNU=35
2190  R      NUIR.KL=FIN.K*SEUI.K
2200  R      NUICR.KL=DELAY3(NUIR.JK,NCT.K)
2210  A      NCT.K=CLIP(NCT2,NCT1,TIME.K,PYEAR)
2220  C      NCT1=10
2230  C      NCT2=10
2240  A      FIN.K=CLIP(FIN2.K,FIN1.K,TIME.K,NMYEAR)
2250  A      FIN1.K=TABHL(FIN1T,FNCR.K,0,2.5,.5)
2260  T      FIN1T=0/.25/.5/.72/.87/1
2270  A      FIN2.K=TABHL(FIN2T,FNCR.K,0,2.5,.5)
2280  T      FIN2T=0/0/0/0/0/0
2290  C      NMYEAR=4000
2300  A      FNCR.K=AFEC.K/NELC.K
2310  A      AFEC.K=(CEC.K*CEG.K+OGEC.K*OGEG.K)/(CEG.K+OGEG.K)
2320  A      NCCR.K=TABHL(NCCRT,TIME.K,1950,2010,10)*1E6
2330  T      NCCRT=1E-6/1E-6/.062/.056/.051/.051/.051
2340  A      NFC.K=TABHL(NFCT,TIME.K,1950,2010,10)*1E-6
2350  T      NFCT=.44/.44/.44/.49/.78/1.23/1.95
2360  A      NELC.K=UCCAF/(NCCR.K*CFCC.K)+NFC.K
2370  C      UCCAF=.14
```

```
2380 A      CFCC.K=CLIP(CFCC2,CFCC1,TIME.K,PYEAR)
2390 C      CFCC1=.7
2400 C      CFCC2=.7
2410 A      NEGC.K=NCCR.K*NUC.K
2420 A      NEG.K=SECF.K*NEGC.K
2430 NOTE
2440 NOTE   ELECTRICITY GENERATION FROM COAL
2450 NOTE
2460 L      CUC.K=CUC.J+(DT)(CUICR.JK-CUDR.JK)
2470 N      CUC=FFUCI*FFUCCI
2480 C      FFUCI=23.8E9
2490 C      FFUCCI=.66
2500 R      CUDR.KL=CUC.K/ALFFU
2510 C      ALFFU=35
2520 R      CUIR.KL=FICU.K*IFFU.K
2530 R      CUICR.KL=DELAY3(CUIR.JK,FFUCT)
2540 C      FFUCT=5
2550 A      IFFU.K=(1-FIN.K)*SEUI.K
2560 A      FICU.K=CLIP(FICU2.K,FICU1.K,TIME.K,PYEAR)
2570 A      FICU1.K=TABHL(FICU1T,FCR.K,0,2,.5)*IMES.K
2580 T      FICU1T=0/.15/.66/.88/1
2590 A      FICU2.K=TABHL(FICU2T,FCR.K,0,2,.5)*IMES.K
2600 T      FICU2T=0/.15/.66/.88/1
2610 A      FCR.K=OGUFC.K/CUFC.K
2620 A      CUFC.K=CPRICE.K/FFECE.K
2630 A      FFECE.K=TABHL(FFECET,TIME.K,1950,2010,10)
2640 T      FFECET=.24/.31/.32/.33/.34/.35/.36
2650 A      FFCCR.K=TABHL(FFCCRT,TIME.K,1950,2010,10)*1E6
2660 T      FFCCRT=.064/.082/.103/.084/.084/.084/.084
2670 A      CEC.K=UCCAF/(CFCC.K*FFCCR.K)+CUFC.K
2680 A      SO2E.K=CEG.K*SO2EF.K
2690 A      SO2EF.K=TABHL(SO2EFT,TIME.K,1950,2010,10)*1E-9
2700 T      SO2EFT=6.1/6.1/5/2/2/2/2
2710 A      SO2ESD.K=CLIP(SO2ESD2,SO2ESD1,TIME.K,PYEAR)
2720 C      SO2ESD1=6.9E6
2730 C      SO2ESD2=6.9E6
2740 A      IMES.K=TABHL(IMEST,SO2ESD.K/SO2E.K,0,1.2,.2)
2750 T      IMEST=0/.04/.14/.38/.92/1/1
2760 A      CEGC.K=CUC.K*FFCCR.K
2770 A      CEG.K=SECF.K*CEGC.K*CPDR.K
2780 A      CDU.K=SECF.K*CEGC.K/FFECE.K
2790 NOTE
2800 NOTE   ELECTRICITY GENERATION FROM OIL AND GAS
2810 NOTE
2820 L      OGUC.K=OGUC.J+(DT)(OGUICR.JK-OGUDR.JK)
2830 N      OGUC=FFUCI*(1-FFUCCI)
2840 R      OGUDR.KL=OGUC.K/ALFFU
2850 R      OGUIR.KL=IFFU.K*(1-FICU.K)
2860 R      OGUICR.KL=DELAY3(OGUIR.JK,FFUCT)
2870 A      OGUFC.K=AOGP.K*OGUFCF/FFECE.K
2880 C      OGUFCF=.54
2890 A      OGEC.K=UCCAF/(CFCC.K*FFCCR.K)+OGUFC.K
2900 A      OGEGC.K=FFCCR.K*OGUC.K
2910 A      OGEG.K=SECF.K*OGEGC.K*OGCDR.K
2920 A      OGDU.K=SECF.K*OGEGC.K/FFECE.K
2930 NOTE
2940 NOTE   *** COAL SECTOR ***
2950 NOTE
2960 NOTE   COAL SUPPLY-DEMAND BALANCE
2970 NOTE
2980 A      CD.K=DCUD.K+CDS.K+CDU.K+CED.K
2990 A      CED.K=TABHL(CEDT,TIME.K,1950,2010,10)*1E15
```

```
3000 T      CEDT=1.7/1.2/2/2/2/2/2
3010 A      CCUF.K=TABHL(CCUFT,CD.K/CPC.K,0,1.2,.1)
3020 T      CCUFT=0/.1/.2/.3/.4/.5/.6/.7/.8/.9/.95/.98/1
3030 A      CPC.K=SCPC.K+UCPC.K
3040 A      CPR.K=SCPR.K+UCPR.K
3050 A      CPDR.K=CPR.K/CD.K
3060 NOTE
3070 NOTE   COAL FINANCING
3080 NOTE
3090 A      CPRICE.K=CPRAT.K*ACC.K
3100 A      ACC.K=(SCPR.K*SCC.K+UCPR.K*UCC.K)/CPR.K
3110 A      CPRAT.K=CLIP(CPRAT2.K,CPRAT1.K,TIME.K,PYEAR)
3120 A      CPRAT1.K=TABHL(CPRAT1T,CD.K/CPC.K,.4,1.6,.2)
3130 T      CPRAT1T=1/1.015/1.13/1.5/1.83/1.97/2
3140 A      CPRAT2.K=TABHL(CPRAT2T,CD.K/CPC.K,.4,1.6,.2)
3150 T      CPRAT2T=1/1.015/1.13/1.5/1.83/1.97/2
3160 A      CROI.K=(CPRICE.K-ACC.K)*CPR.K/CC.K
3170 A      CC.K=SC.K+UC.K
3180 A      ACROI.K=SMOOTH(CROI.K,CROIAT)
3190 C      CROIAT=5
3200 A      FCRI.K=CLIP(FCRI2.K,FCRI1.K,TIME.K,PYEAR)
3210 A      FCRI1.K=TABHL(FCRI1T,ACROI.K,0,.25,.05)
3220 T      FCRI1T=0/.03/.13/.2/.23/.25
3230 A      FCRI2.K=TABHL(FCRI2T,ACROI.K,0,.25,.05)
3240 T      FCRI2T=0/.03/.13/.2/.23/.25
3250 A      CREV.K=CPRICE.K*CPR.K
3260 A      CCINV.K=FCRI.K*CREV.K
3270 NOTE
3280 NOTE   UNDERGROUND CAPITAL
3290 NOTE
3300 L      UC.K=UC.J+(DT)(UCICR.JK-UCDR.JK)
3310 N      UC=UCI
3320 C      UCI=2.5E9
3330 R      UCDR.KL=UC.K/ALCC
3340 C      ALCC=20
3350 R      UCIR.KL=FIU.K*CCINV.K
3360 R      UCICR.KL=DELAY3(UCIR.JK,UMCT)
3370 C      UMCT=5
3380 A      FIU.K=CLIP(FIU2.K,FIU1.K,TIME.K,PYEAR)
3390 A      FIU1.K=TABHL(FIU1T,SCC.K/UCC.K,0,2.5,.5)
3400 T      FIU1T=0/.3/.75/.95/.99/1
3410 A      FIU2.K=TABHL(FIU2T,SCC.K/UCC.K,0,2.5,.5)
3420 T      FIU2T=0/.3/.75/.95/.99/1
3430 NOTE
3440 NOTE   UNDERGROUND LABOR SUPPLY
3450 NOTE
3460 L      UCLS.K=UCLS.J+(DT)(NHR.JK)
3470 N      UCLS=UCLSI
3480 C      UCLSI=3.73E5
3490 R      NHR.KL=CLIP(HR.K,LR.K,HR.K,0)
3500 A      LR.K=(RCLS.K-UCLS.K)/LAT
3510 C      LAT=.5
3520 A      HR.K=(RCLS.K-UCLS.K)/HAT.K
3530 A      RCLS.K=UC.K/RCLR.K
3540 A      RCLR.K=TABLE(RCLRT,TIME.K,1950,2010,10)*1E4
3550 T      RCLRT=.67/1.7/3.1/4.4/5.7/6.9/8.2
3560 A      HAT.K=TABHL(HATT,PAR.K/(UW.K*HAWR),0,1.2,.2)
3570 T      HATT=.5/.8/2/4/6.5/10/14
3580 C      HAWR=.14E-3
3590 A      UW.K=CLIP(UWP,UWH,TIME.K,PYEAR)
3600 C      UWH=7000
3610 C      UWP=7000
3620 A      PAR.K=DLINF3(AR.K,SPDT)
```

```
3630 C      SPDT=10
3640 A      AR.K=SMOOTH(SS.K,SIDT)
3650 C      SIDT=2
3660 A      SS.K=CLIP(SSN.K,SSH,TIME.K,1970)
3670 C      SSH=1
3680 A      SSN.K=CLIP(SSP,SS69,TIME.K,PYEAR)
3690 C      SS69=.5
3700 C      SSP=.5
3710 NOTE
3720 NOTE   UNDERGROUND PRODUCTION
3730 NOTE
3740 A      UCPC.K=UCPC70*(LABR.K^LE)*(CAPR.K^CE)*PMS.K*UPMD.K
3750 C      UCPC70=10.6E15
3760 C      LE=.53
3770 C      CE=.47
3780 A      LABR.K=UCLS.K/UCLS70
3790 C      UCLS70=108000
3800 A      CAPR.K=UC.K/UC70
3810 C      UC70=3.4E9
3820 A      PMS.K=TABHL(PMST,AR.K/ARH,0,1.4,.2)
3830 T      PMST=0/.41/.68/.82/.93/1/1.05/1.08
3840 C      ARH=1
3850 A      UPMD.K=TABLE(UPMDT,FUCRR.K,0,1,.2)
3860 T      UPMDT=0/.1/.22/.45/.9/1
3870 A      FUCRR.K=UCR.K/UCRI
3880 L      UCR.K=UCR.J+(DT)(-UCDPLR.JK)
3890 N      UCR=UCRI
3900 C      UCRI=15E18
3910 R      UCDPLR.KL=UCPR.K
3920 A      UCPR.K=CCUF.K*UCPC.K
3930 A      UCOC.K=UCAF*UC.K/(CCUFN*UCPC.K)
3940 C      UCAF=.65
3950 C      CCUFN=.8
3960 A      ULC.K=UW.K*UCLS.K/UCPR.K
3970 A      UCC.K=ULC.K+UCOC.K
3980 NOTE
3990 NOTE   SURFACE COAL PRODUCTION
4000 NOTE
4010 L      SC.K=SC.J+(DT)(SCICR.JK-SCDR.JK)
4020 N      SC=SCI
4030 C      SCI=1.4E9
4040 R      SCDR.KL=SC.K/ALCC
4050 R      SCIR.KL=(1-FIU.K)*CCINV.K
4060 R      SCICR.KL=DELAY3(SCIR.JK,SMCT)
4070 C      SMCT=3
4080 A      SCPC.K=SCCR.K*SC.K
4090 A      SCCR.K=SCCRN*SPMD.K
4100 C      SCCRN=3.5E6
4110 A      SPMD.K=TABLE(SPMDT,FSCRR.K,0,1,.2)
4120 T      SPMDT=0/.13/.35/.7/1/1
4130 A      FSCRR.K=SCR.K/SCRI
4140 L      SCR.K=SCR.J+(DT)(-SCDPLR.JK)
4150 N      SCR=SCRI
4160 C      SCRI=2.4E18
4170 R      SCDPLR.KL=SCPR.K
4180 A      SCPR.K=CCUF.K*SCPC.K
4190 A      SCC.K=SCAF/(SCCR.K*CCUFN)+SCP.K
4200 C      SCAF=.75
4210 A      SCP.K=CLIP(SCP2,SCP1,TIME.K,PYEAR)
4220 C      SCP1=0
4230 C      SCP2=0
4240 NOTE
4250 NOTE   GROSS ENERGY INPUTS
```

```
4260 NOTE
4270 A      GED.K=GOGD.K-SOGPC.K*DOGCUF.K+CD.K+GELI.K
4280 A      GELI.K=3.12*NEG.K+3.08*HG.K
4290 A      GEC.K=COGPR.K+OGI.K+CPR.K+GELI.K
4300 A      GECDR.K=GEC.K/GED.K
4310 A      COGCR.K=CPR.K+COGPR.K
4320 A      COGNCR.K=COGCR.K+GELI.K
4330 NOTE
4340 NOTE   SUPPLEMENTARY VARIABLES
4350 NOTE
4360 A      COGEG.K=OGEG.K+CEG.K
4370 A      COGNEG.K=COGEG.K+NEG.K
4380 A      TEI.K=DOGINV.K+SEUI.K+CCINV.K
4390 A      EIGNPR.K=TEI.K/GNP.K
4400 A      COGD.K=DCUD.K+NOGD.K
4410 NOTE
4420 NOTE   CONTROL CARDS
4430 NOTE
4440 N      TIME=1950
4450 C      PYEAR=1977
4460 SPEC   DT=.5/LENGTH=2010/PLTPER=2/PRTPER=0
4470 PLOT   CD=D,CPR=P,UCPR=U(0,2E17)/CPDR=*(0,1)
4480 PLOT   TEG=G,NELD=D,COGNEG=N,COGEG=C,OGEG=O(0,60E15)/EGDR=*(0,1)
4490 PLOT   GOGD=D,OGC=C,COGPR=O,DOGPR=P(0,1.2E17)
4500 PLOT   GED=G,NED=D,DCUD=C,COGD=O(0,2.4E17)
4510 PLOT   AEP=$,CPRICE=+,AOGP=#,EP=*(-4E-6,12E-6)/EIGNPR=%(0,.8)
4520 PLOT   GEC=C,GED=D,COGPR=O,COGCR=C,COGNCR=N(0,2.4E17)/GECDR=*(0,1)
4530 NOTE
4540 NOTE   PARAMETER CHANGES FOR THE COAL2 POLICY RUNS
4550 NOTE
4560 NOTE   REFERENCE RUN
4570 NOTE
4580 RUN    REFERENCE RUN
4590 NOTE
4600 NOTE   ENERGY POLICY ANALYSIS
4610 NOTE
4620 NOTE   DEMAND POLICIES
4630 NOTE
4640 T      IDMP2T=1.2/1/.76/.64/.57/.53/.49/.46/.44/.42/.40
4650 RUN    ACCELERATED CONSERVATION PROJECTION
4660 T      IDMP2T=1.2/1/.76/.64/.57/.53/.49/.46/.44/.42/.40
4670 T      LTGRT=3.55/3.55/3.55/3.4/2.6/1.2/0
4680 RUN    ZERO ENERGY GROWTH PROJECTION
4690 NOTE
4700 NOTE   SUPPLY POLICIES
4710 NOTE
4720 C      NCT2=6
4730 T      NFCT=.44/.44/.44/.49/.64/.82/1.07
4740 RUN    ACCELERATED NUCLEAR PROJECTION
4750 CP     PPRAT=2.07
4760 CP     SDT=4
4770 TP     FIASS2T=0/.2/.5/.8/.9/.95/.98/.99/1
4780 CP     RRT=1977
4790 TP     SECFT=0/.2/.4/.6/.8/.85
4800 CP     SECF2=.60
4810 CP     CFCC2=.75
4820 TP     SO2EFT=6.1/6.1/5/2/.6/.6/.6
4830 TP     FFCCRT=.064/.082/.103/.084/.073/.073/.073
4840 CP     SCP2=.2E-6
4850 TP     SPMDT=0/.11/.32/.67/1/1
4860 TP     CPRAT2T=1.3/1.3/1.3/1.5/1.83/1.97/2
4870 RUN    ACCELERATED COAL PROJECTION
4880 NOTE
```

```
4890 NOTE   COMBINED SUPPLY AND DEMAND POLICIES
4900 NOTE
4910 TP     IDMP2T=1.2/1/.76/.64/.57/.53/.49/.46/.44/.42/.40
4920 RUN    ACCELERATED COAL AND CONSERVATION
4930 TP     LTGRT=3.55/3.55/3.55/3.4/2.6/1.2/0
4940 RUN    ZEG AND ACCELERATED COAL PROJECTION
```

Appendix C
Definition File

1000	ACC	AVERAGE COAL COST (1970 DOLLARS/BTU)
1010	ACROI	AVERAGE COAL RETURN ON INVESTMENT (FRACTION/YEAR)
1020	AECOST	AVERAGE ELECTRICITY COST (1970 DOLLARS/BTU)
1030	AEP	AVERAGE ELECTRICITY PRICE (1970 DOLLARS/BTU)
1040	AEPN	AVERAGE ENERGY PRICE NORMAL (1970 DOLLARS/BTU)
1050	AEROI	AVERAGE ELECTRICITY RETURN ON INVESTMENT (FRACTION/YEAR)
1060	AFEC	AVERAGE FOSSIL ELECTRICITY COST (1970 DOLLARS/BTU)
1070	ALCC	AVERAGE LIFETIME OF COAL CAPITAL (YEARS)
1080	ALELP	ALLOWED ELECTRICITY PRICE (1970 DOLLARS/BTU)
1090	ALFFU	AVERAGE LIFETIME OF FOSSIL-FIRED UTILITIES (YEARS)
1100	ALNU	AVERAGE LIFETIME OF NUCLEAR UTILITIES (YEARS)
1110	ALOGC	AVERAGE LIFETIME OF OIL AND GAS CAPITAL (YEARS)
1120	ALSYNC	AVERAGE LIFETIME OF SYNTHETIC CAPITAL (YEARS)
1130	AOGP	AVERAGE OIL AND GAS PRICE (1970 DOLLARS/BTU)
1140	AOGROI	AVERAGE OIL AND GAS RETURN ON INVESTMENT (FRACTION/YEAR)
1150	AR	ACCIDENT RATE (FATALITIES/MILLION MAN-HOURS)
1160	ARH	ACCIDENT RATE HISTORICAL (FATALITIES/MILLION MAN-HOURS)
1170	AROR	ALLOWED RATE OF RETURN (FRACTION/YEAR)
1180	CAPR	CAPITAL RATIO (DIMENSIONLESS)
1190	CC	COAL CAPITAL (1970 DOLLARS)
1200	CCCRI	CONVENTIONAL CAPACITY/CAPITAL RATIO INITIAL (BTU/DOLLAR-YEAR)
1210	CCINV	COAL CAPITAL INVESTMENT (1970 DOLLARS/YEAR)
1220	CCOST	CONVENTIONAL OIL AND GAS COST (1970 DOLLARS/BTU)
1230	CCUF	COAL CAPACITY UTILIZATION FACTOR (FRACTION)
1240	CCUFN	COAL CAPACITY UTILIZATION FACTOR NORMAL (FRACTION)
1250	CCUFT	CCUF TABLE
1260	CD	COAL DEMAND (BTU/YEAR)
1270	CDS	COAL DEMANDED FROM SYNTHETICS (BTU/YEAR)
1280	CDSM	COAL DEMAND SUBSTITUTION MULTIPLIER (DIMENSIONLESS)
1290	CDSMT	CDSM TABLE
1300	CDU	COAL DEMANDED BY UTILITIES (BTU/YEAR)
1310	CE	CAPITAL ELASTICITY (DIMENSIONLESS)
1320	CEC	COAL ELECTRICITY COST (1970 DOLLARS/BTU)
1330	CED	COAL EXPORT DEMAND (BTU/YEAR)
1340	CEDT	CED TABLE
1350	CEG	COAL ELECTRICITY GENERATION (BTU/YEAR)
1360	CEGC	COAL ELECTRICITY GENERATION CAPACITY (BTU/YEAR)
1370	CFCC	CAPACITY FACTOR FOR COST CALCULATIONS (FRACTION)
1380	CFCC1	VALUE OF CFCC BEFORE TIME=PYEAR (FRACTION)
1390	CFCC2	VALUE OF CFCC AFTER TIME=PYEAR (FRACTION)
1400	COGC	CONVENTIONAL OIL AND GAS CAPITAL (1970 DOLLARS)
1410	COGCCR	CONVENTIONAL OIL AND GAS CAPACITY/CAPITAL RATIO (BTU/DOLLAR-YEAR)
1420	COGCI	CONVENTIONAL OIL AND GAS CAPITAL INITIAL (1970 DOLLARS)
1430	COGCR	COAL, OIL AND GAS CONSUMPTION RATE (BTU/YEAR)
1440	COGCT	CONVENTIONAL OIL AND GAS CONSTRUCTION TIME (YEARS)

1450 COGD	COAL,OIL, AND GAS DEMAND (BTU/YEAR)
1460 COGDPL	CONVENTIONAL OIL AND GAS DEPLETION RATE (BTU/YEAR)
1470 COGDR	CONVENTIONAL OIL AND GAS DEPRECIATION RATE (1970 DOLLARS/YEAR)
1480 COGEG	COAL, OIL AND GAS ELECTRICITY GENERATION (BTU/YEAR)
1490 COGICR	CONVENTIONAL OIL AND GAS INVESTMENT COMPLETION RATE (1970 DOLLARS/Y
1500 COGIR	CONVENTIONAL OIL AND GAS INVESTMENT RATE (1970 DOLLARS/YEAR)
1510 COGNCR	COAL,OIL,GAS, AND NUCLEAR CONSUMPTION RATE (BTU/YEAR)
1520 COGNEG	COAL, OIL, GAS AND NUCLEAR ELECTRICITY GENERATION (BTU/YEAR)
1530 COGP50	CUMULATIVE OIL AND GAS PRODUCTION IN 1950 (BTU)
1540 COGPC	CONVENTIONAL OIL AND GAS PRODUCTION CAPACITY (BTU/YEAR)
1550 COGPR	CONVENTIONAL OIL AND GAS PRODUCTION RATE (BTU/YEAR)
1560 COGR	CONVENTIONAL OIL AND GAS RESOURCES (BTU)
1570 COGRI	CONVENTIONAL OIL AND GAS RESOURCES INITIAL (BTU)
1580 COPR	COAL/OIL PROFIT RATIO (DIMENSIONLESS)
1590 CPC	COAL PRODUCTION CAPACITY (BTU/YEAR)
1600 CPDR	COAL PRODUCTION/DEMAND RATIO (DIMENSIONLESS)
1610 CPR	COAL PRODUCTION RATE (BTU/YEAR)
1620 CPRAT	COAL PRICE RATIO (DIMENSIONLESS)
1630 CPRAT1	VALUE OF CPRAT BEFORE TIME=PYEAR (DIMENSIONLESS)
1640 CPRAT1T	CPRAT1 TABLE
1650 CPRAT2	VALUE OF CPRAT AFTER TIME=PYEAR (DIMENSIONLESS)
1660 CPRAT2T	CPRAT2 TABLE
1670 CPRICE	COAL PRICE (1970 DOLLARS/BTU)
1680 CREV	COAL REVENUES (1970 DOLLARS/YEAR)
1690 CROI	COAL RETURN ON INVESTMENT (FRACTION/YEAR)
1700 CROIAT	COAL RETURN ON INVESTMENT AVERAGING TIME (YEARS)
1710 CUC	COAL UTILITIES CAPITAL (1970 DOLLARS)
1720 CUDR	COAL UTILITIES DEPRECIATION RATE (1970 DOLLARS/YEAR)
1730 CUFC	COAL UTILITIES FUEL COST (1970 DOLLARS/BTU)
1740 CUICR	COAL UTILITIES INVESTMENT COMPLETION RATE (1970 DOLLARS/YEAR)
1750 CUIR	COAL UTILITIES INVESTMENT RATE (1970 DOLLARS/YEAR)
1760 DAT	DEMAND ADJUSTMENT TIME (YEARS)
1770 DCOST	DOMESTIC COST (1970 DOLLARS/BTU)
1780 DCUD	DIRECT COAL USE DEMAND (BTU/YEAR)
1790 DEREGT	TIME OF OIL AND GAS PRICE DEREGULATION (YEAR)
1800 DMP	DEMAND MULTIPLIER FROM PRICE (DIMENSIONLESS)
1810 DOGCUF	DOMESTIC OIL AND GAS CAPACITY UTILIZATION FACTOR (FRACTION)
1820 DOGCUFT	DOGCUF TABLE
1830 DOGINV	DOMESTIC OIL AND GAS INVESTMENT (1970 DOLLARS/YEAR)
1840 DOGP	DOMESTIC OIL AND GAS PRICE (1970 DOLLARS/BTU)
1850 DOGPC	DOMESTIC OIL AND GAS PRODUCTION CAPACITY (BTU/YEAR)
1860 DOGPR	DOMESTIC OIL AND GAS PRODUCTION RATE (BTU/YEAR)
1870 DOGREV	DOMESTIC OIL AND GAS REVENUES (1970 DOLLARS/YEAR)
1880 DOGROI	DOMESTIC OIL AND GAS RETURN ON INVESTMENT (FRACTION/YEAR)
1890 ECOPR	EFFECTIVE COAL/OIL PRICE RATIO (DIMENSIONLESS)
1900 ECOPR70	VALUE OF ECOPR IN 1970 (DIMENSIONLESS)
1910 EDSM	ELECTRICITY DEMAND SUBSTITUTION MULTIPLIER (DIMENSIONLESS)
1920 EDSMT	EDSM TABLE
1930 EEPR	EFFECTIVE ELECTRICITY PRICE RATIO (DIMENSIONLESS)
1940 EEPR70	VALUE OF EEPR IN 1970 (DIMENSIONLESS)
1950 EGDR	ELECTRICITY GENERATION/DEMAND RATIO (DIMENSIONLESS)
1960 EGNPR70	ENERGY/GNP RATIO IN 1970 (BTU/DOLLAR)
1970 EIGNPR	ENERGY INVESTMENT/GNP RATIO (FRACTION)
1980 EMYEAR	EMBARGO YEAR (YEAR)
1990 EP	ELECTRICITY PRICE (1970 DOLLARS/BTU)
2000 EPR	ELECTRICITY PRICE RATIO (DIMENSIONLESS)
2010 EROI	ELECTRICITY RETURN ON INVESTMENT (FRACTION/YEAR)
2020 EROIAT	ELECTRICITY RETURN ON INVESTMENT AVERAGING TIME (YEARS)
2030 EROIM	ELECTRICITY RETURN ON INVESTMENT MULTIPLIER (DIMENSIONLESS)
2040 EROIMT	EROIM TABLE
2050 FCR	FUEL COST RATIO (DIMENSIONLESS)
2060 FCRI	FRACTION OF COAL REVENUES INVESTED (FRACTION)
2070 FCRI1	VALUE OF FCRI BEFORE TIME=PYEAR (FRACTION)
2080 FCRI1T	FCRI1 TABLE
2090 FCRI2	VALUE OF FCRI AFTER TIME=PYEAR (FRACTION)
2100 FCRI2T	FCRI2 TABLE
2110 FEDC	FRACTION OF ENERGY DEMANDED AS COAL (FRACTION)
2120 FEDCT	FEDC TABLE
2130 FEDE	FRACTION OF ENERGY DEMANDED AS ELECTRICITY (FRACTION)
2140 FEDET	FEDE TABLE
2150 FFCCR	FOSSIL-FIRED CAPACITY/CAPITAL RATIO (BTU/DOLLAR-YEAR)
2160 FFCCRT	FFCCR TABLE
2170 FFECE	FOSSIL-FIRED ELECTRICITY CONVERSION EFFICIENCY (FRACTION)

```
2180 FFECET    FFECE TABLE
2190 FFUCCI    FRACTION OF FOSSIL UTILITIES CAPITAL IN COAL INITIAL (FRACTION)
2200 FFUCI     FOSSIL-FIRED UTILITIES CAPITAL INITIAL (1970 DOLLARS)
2210 FFUCT     FOSSIL-FIRED UTILITIES CONSTRUCTION TIME (YEARS)
2220 FIASS     FRACTION OF INVESTMENT ALLOCATED TO SYNTHETICS (FRACTION)
2230 FIASS1    VALUE OF FIASS BEFORE TIME=PYEAR (FRACTION)
2240 FIASS1T   FIASS1 TABLE
2250 FIASS2    VALUE OF FIASS AFTER TIME=PYEAR (FRACTION)
2260 FIASS2T   FIASS2 TABLE
2270 FICU      FRACTION INVESTED IN COAL UTILITIES (FRACTION)
2280 FICU1     VALUE OF FICU BEFORE TIME=PYEAR (FRACTION)
2290 FICU1T    FICU1 TABLE
2300 FICU2     VALUE OF FICU AFTER TIME=PYEAR (FRACTION)
2310 FICU2T    FICU2 TABLE
2320 FIN       FRACTION INVESTED IN NUCLEAR (FRACTION)
2330 FIN1      VALUE OF FIN BEFORE TIME=NMYEAR (FRACTION)
2340 FIN1T     FIN1 TABLE
2350 FIN2      VALUE OF FIN AFTER TIME=NMYEAR (FRACTION)
2360 FIN2T     FIN2 TABLE
2370 FIU       FRACTION INVESTED IN UNDERGROUND (FRACTION)
2380 FIU1      VALUE OF FIU BEFORE TIME=PYEAR (FRACTION)
2390 FIU1T     FIU1 TABLE
2400 FIU2      VALUE OF FIU AFTER TIME=PYEAR (FRACTION)
2410 FIU2T     FIU2 TABLE
2420 FNCR      FOSSIL/NUCLEAR COST RATIO (DIMENSIONLESS)
2430 FOGRR     FRACTION OF OIL AND GAS RESOURCES REMAINING (FRACTION)
2440 FOP2      VALUE OF OGIP AFTER TIME=TYEAR (1970 DOLLARS/BTU)
2450 FOPE      FOREIGN OIL PRICE AFTER 1973 EMBARGO (1970 DOLLARS/BTU)
2460 FOPH      FOREIGN OIL PRICE HISTORICAL (1970 DOLLARS/BTU)
2470 FOPT      VALUE OF OGIP AFTER TIME=TYEAR (1970 DOLLARS/BTU)
2480 FSCRR     FRACTION OF SURFACE COAL RESOURCES REMAINING (FRACTION)
2490 FSERI     FRACTION OF STEAM-ELECTRIC REVENUES INVESTED (FRACTION)
2500 FSERIT    FSERI TABLE
2510 FUCRR     FRACTION OF UNDERGROUND COAL RESOURCES REMAINING (FRACTION)
2520 GDOGD     GROSS DOMESTIC OIL AND GAS DEMAND (BTU/YEAR)
2530 GEC       GROSS ENERGY CONSUMPTION (BTU/YEAR)
2540 GECDR     GROSS ENERGY CONSUMPTION/DEMAND RATIO (DIMENSIONLESS)
2550 GED       GROSS ENERGY DEMAND (BTU/YEAR)
2560 GELI      GROSS ELECTRICITY INPUTS (BTU/YEAR)
2570 GNP       GROSS NATIONAL PRODUCT (1970 DOLLARS/YEAR)
2580 GNP70     VALUE OF GNP IN 1970 (1970 DOLLARS/YEAR)
2590 GNPGR     GNP GROWTH RATE (PERCENT/YEAR)
2600 GNPGR1    GNP GROWTH RATE AFTER 1973 (PERCENT/YEAR)
2610 GNPGRT    GNPGR TABLE
2620 GNPI      GROSS NATIONAL PRODUCT INITIAL (1970 DOLLARS/YEAR)
2630 GNPIR     GNP INCREASE RATE (1970 DOLLARS/YEAR-YEAR)
2640 GOGC      GROSS OIL AND GAS CONSUMPTION (BTU/YEAR)
2650 GOGD      GROSS OIL AND GAS DEMAND (BTU/YEAR)
2660 HAT       HIRING ADJUSTMENT TIME (YEARS)
2670 HATT      HAT TABLE
2680 HAWR      HISTORICAL ACCIDENT/WAGE RATIO (FATALITY-YEARS/MILLION HOURS-DOLLAR)
2690 HG        HYDROPOWER GENERATION (BTU/YEAR)
2700 HGT       HG TABLE
2710 HPRAT     HISTORICAL VALUE OF REGULATED PROFIT RATIO (DIMENSIONLESS)
2720 HR        HIRING RATE (MEN/YEAR)
2730 IAROR     AROR AFTER TIME=RRT (FRACTION/YEAR)
2740 IDMP      INDICATED DEMAND MULTIPLIER FROM PRICE (DIMENSIONLESS)
2750 IDMP1     VALUE OF IDMP BEFORE TIME=PYEAR (DIMENSIONLESS)
2760 IDMP1T    IDMP1 TABLE
2770 IDMP2     VALUE OF IDMP AFTER TIME=PYEAR (DIMENSIONLESS)
2780 IDMP2T    IDMP2 TABLE
2790 IFFU      INVESTMENT IN FOSSIL-FIRED UTILITIES (1970 DOLLARS/YEAR)
2800 IFIASS    INDICATED FRACTION OF INVESTMENT ALLOCATED TO SYNTHETICS (FRACTION)
2810 IMES      INVESTMENT MULTIPLIER FROM EMISSIONS STANDARD (DIMENSIONLESS)
2820 IMEST     IMES TABLE
2830 LABR      LABOR RATIO (DIMENSIONLESS)
2840 LAT       LAYOFF ADJUSTMENT TIME (YEARS)
2850 LE        LABOR ELASTICITY (DIMENSIONLESS)
2860 LR        LAYOFF RATE (MEN/YEAR)
2870 LTGR      LONG-TERM GROWTH RATE (PERCENT/YEAR)
2880 LTGRT     LTGR TABLE
2890 NAROR     AROR BEFORE TIME=RRT (FRACTION/YEAR)
2900 NCCR      NUCLEAR CAPACITY/CAPITAL RATIO (BTU/DOLLAR-YEAR)
```

```
2910 NCCRT    NCCR TABLE
2920 NCDR     NUCLEAR CAPITAL DEPRECIATION RATE (1970 DOLLARS/YEAR)
2930 NCT      NUCLEAR CONSTRUCTION TIME (YEARS)
2940 NCT1     VALUE OF NCT BEFORE TIME=PYEAR (YEARS)
2950 NCT2     VALUE OF NCT AFTER TIME=PYEAR (YEARS)
2960 NCUF     NORMAL CAPACITY UTILIZATION FACTOR (FRACTION)
2970 NEC      NET ENERGY CONSUMPTION (BTU/YEAR)
2980 NED      NET ENERGY DEMAND (BTU/YEAR)
2990 NEG      NUCLEAR ELECTRICITY GENERATION (BTU/YEAR)
3000 NEGC     NUCLEAR ELECTRICITY GENERATION CAPACITY (BTU/YEAR)
3010 NELC     NUCLEAR ELECTRICITY COST (1970 DOLLARS/BTU)
3020 NELD     NET ELECTRICITY DEMAND (BTU/YEAR)
3030 NFC      NUCLEAR FUEL COST (1970 DOLLARS/BTU)
3040 NFCT     NFC TABLE
3050 NHR      NORMAL HIRING RATE (MEN/YEAR)
3060 NMYEAR   YEAR OF NUCLEAR MORATORIUM (YEAR)
3070 NOGD     NET OIL AND GAS DEMAND (BTU/YEAR)
3080 NUC      NUCLEAR UTILITIES CAPITAL (1970 DOLLARS)
3090 NUCI     NUCLEAR UTILITIES CAPITAL INITIAL (1970 DOLLARS)
3100 NUICR    NUCLEAR UTILITIES INVESTMENT COMPLETION RATE (1970 DOLLARS/YEAR)
3110 NUIR     NUCLEAR UTILITIES INVESTMENT RATE (1970 DOLLARS/YEAR)
3120 OGC      OIL AND GAS CONSUMPTION (BTU/YEAR)
3130 OGCCAF   OIL AND GAS CAPITAL COST ANNUALIZATION FACTOR (FRACTION/YEAR)
3140 OGCDR    OIL AND GAS CONSUMPTION/DEMAND RATIO (DIMENSIONLESS)
3150 OGCEM    OIL AND GAS CAPITAL EFFICIENCY MULTIPLIER (DIMENSIONLESS)
3160 OGCEMT   OGCEM TABLE
3170 OGDU     OIL AND GAS DEMANDED BY UTILITIES (BTU/YEAR)
3180 OGEC     OIL AND GAS ELECTRICITY COST (1970 DOLLARS/BTU)
3190 OGEG     OIL AND GAS ELECTRICITY GENERATION (BTU/YEAR)
3200 OGEGC    OIL AND GAS ELECTRICITY GENERATION CAPACITY (BTU/YEAR)
3210 OGI      OIL AND GAS IMPORTS (BTU/YEAR)
3220 OGIP     OIL AND GAS IMPORTS PRICE (1970 DOLLARS/BTU)
3230 OGUC     OIL AND GAS UTILITIES CAPITAL (1970 DOLLARS)
3240 OGUDR    OIL AND GAS UTILITIES DEPRECIATION RATE (1970 DOLLARS/YEAR)
3250 OGUFC    OIL AND GAS UTILITIES FUEL COST (1970 DOLLARS/YEAR)
3260 OGUFCF   OIL AND GAS UTILITIES FUEL COST FACTOR (DIMENSIONLESS)
3270 OGUICR   OIL AND GAS UTILITIES INVESTMENT COMPLETION RATE (1970 DOLLARS/YEAR)
3280 OGUIR    OIL AND GAS UTILITIES INVESTMENT RATE (1970 DOLLARS/YEAR)
3290 OIQ      OIL IMPORT QUOTA (FRACTION)
3300 PAR      PERCEIVED ACCIDENT RATE (FATALITIES/MILLION MAN-HOURS)
3310 PMS      PRODUCTION MULTIPLIER FROM SAFETY (DIMENSIONLESS)
3320 PMST     PMS TABLE
3330 POGRR    PERCENT OF OIL AND GAS REVENUES REINVESTED (FRACTION)
3340 POGRR1   VALUE OF POGRR BEFORE TIME=PYEAR (FRACTION)
3350 POGRR1T  POGRR1 TABLE
3360 POGRR2   VALUE OF POGRR AFTER TIME=PYEAR (FRACTION)
3370 POGRR2T  POGRR2 TABLE
3380 PPRAT    POLICY VALUE OF REGULATED PROFIT RATIO (DIMENSIONLESS)
3390 PYEAR    YEAR OF POLICY IMPLEMENTATION (YEAR)
3400 RCLR     REQUIRED CAPITAL/LABOR RATIO (DOLLARS/MAN)
3410 RCLRT    RCLR TABLE
3420 RCLS     REQUIRED COAL LABOR SUPPLY (MEN)
3430 RCYEAR   RECOVERY YEAR (YEAR)
3440 REGD     REGULATION DELAY (YEARS)
3450 RELP     REGULATED ELECTRICITY PRICE (1970 DOLLARS/BTU)
3460 RL       REGULATORY LAG (YEARS)
3470 RLT      REGULATORY LAG TIME (YEARS)
3480 ROGI     RESTRICTED OIL AND GAS IMPORTS (BTU/YEAR)
3490 ROGP     REGULATED OIL AND GAS PRICE (1970 DOLLARS/BTU)
3500 ROIAT    RETURN ON INVESTMENT ADJUSTMENT TIME (YEARS)
3510 RPRAT    REGULATED PRICE RATIO (DIMENSIONLESS)
3520 RRT      TIME OF IMPLEMENTATION OF RATE RELIEF POLICY (YEAR)
3530 RSYEAR   RECESSION START YEAR (YEAR)
3540 RYGR     RECESSION YEAR GROWTH RATE (PERCENT/YEAR)
3550 RYGRT    RYGR TABLE
3560 SC       SURFACE CAPITAL (1970 DOLLARS)
3570 SCAF     SURFACE COST ANNUALIZATION FACTOR (FRACTION/YEAR)
3580 SCC      SURFACE COAL COSTS (1970 DOLLARS/BTU)
3590 SCCR     SURFACE COAL CAPACITY/CAPITAL RATIO (BTU/DOLLAR-YEAR)
3600 SCCRN    SURFACE COAL CAPACITY/CAPITAL RATIO NORMAL (BTU/DOLLAR-YEAR)
3610 SCDPLR   SURFACE COAL DEPLETION RATE (BTU/YEAR)
3620 SCDR     SURFACE CAPITAL DEPRECIATION RATE (1970 DOLLARS/YEAR)
3630 SCE      SYNTHETICS CONVERSION EFFICIENCY (FRACTION)
```

```
3640 SCI        SURFACE CAPITAL INITIAL (1970 DOLLARS)
3650 SCICR      SURFACE CAPITAL INVESTMENT COMPLETION RATE (1970 DOLLARS/YEAR)
3660 SCIR       SURFACE CAPITAL INVESTMENT RATE (1970 DOLLARS/YEAR)
3670 SCOPR      SMOOTHED COAL-OIL PRICE RATIO (DIMENSIONLESS)
3680 SCOPR70    VALUE OF SCOPR IN 1970 (DIMENSIONLESS)
3690 SCOST      SYNTHETIC OIL AND GAS COST (1970 DOLLARS/BTU)
3700 SCP        SURFACE COST FROM POLICY (1970 DOLLARS/BTU)
3710 SCP1       VALUE OF SCP BEFORE TIME=PBTU (1970 DOLLARS/BTU)
3720 SCP2       VALUE OF SCP AFTER TIME=PBTU (1970 DOLLARS/BTU)
3730 SCPC       SURFACE COAL PRODUCTION CAPACITY (BTU/YEAR)
3740 SCPR       SURFACE COAL PRODUCTION RATE (BTU/YEAR)
3750 SCR        SURFACE COAL RESOURCES (BTU)
3760 SCRI       SURFACE COAL RESOURCES INITIAL (BTU)
3770 SCT        SYNTHETIC CONSTRUCTION TIME (YEARS)
3780 SDT        SYNTHETIC DEVELOPMENT TIME (YEARS)
3790 SEC        STEAM-ELECTRIC CAPITAL (1970 DOLLARS)
3800 SECF       STEAM-ELECTRIC CAPACITY FACTOR (FRACTION)
3810 SECF1      VALUE OF SECFN BEFORE TIME=PYEAR (FRACTION)
3820 SECF2      VALUE OF SECFN AFTER TIME=PYEAR (FRACTION)
3830 SECFN      STEAM-ELECTRIC CAPACITY FACTOR NORMAL (FRACTION)
3840 SECFT      SECF TABLE
3850 SED        STEAM-ELECTRIC DEMAND (BTU/YEAR)
3860 SEG        STEAM-ELECTRIC GENERATION (BTU/YEAR)
3870 SEGC       STEAM-ELECTRIC GENERATION CAPACITY (BTU/YEAR)
3880 SEPR       SMOOTHED ELECTRICITY-OIL PRICE RATIO (DIMENSIONLESS)
3890 SEPR70     VALUE OF SEPR IN 1970 (DIMENSIONLESS)
3900 SER        STEAM-ELECTRIC REVENUES (1970 DOLLARS/YEAR)
3910 SEUI       STEAM-ELECTRIC UTILITIES INVESTMENT (1970 DOLLARS/YEAR)
3920 SFC        SYNTHETICS FUEL COST (1970 DOLLARS/BTU)
3930 SID        SAFETY IMPLEMENTATION DELAY (YEARS)
3940 SIDT       SAFETY IMPLEMENTATION DELAY TIME (YEARS)
3950 SMCT       SURFACE MINE CONSTRUCTION TIME (YEARS)
3960 SO2E       SO2 EMISSIONS (TONS/YEAR)
3970 SO2EF      SO2 EMISSIONS FACTOR (TONS/BTU)
3980 SO2EFT     SO2EF TABLE
3990 SO2ESD     SO2 EMISSION STANDARD (TONS/YEAR)
4000 SO2ESD1    VALUE OF SO2ESD BEFORE TIME=PYEAR (TONS/YEAR)
4010 SO2ESD2    VALUE OF SO2ESD AFTER TIME=PYEAR (TONS/YEAR)
4020 SOGC       SYNTHETIC OIL AND GAS CAPITAL (1970 DOLLARS)
4030 SOGCDR     SYNTHETIC OIL AND GAS CAPITAL DEPRECIATION RATE (1970 DOLLARS/YEAR)
4040 SOGCI      SYNTHETIC OIL AND GAS CAPITAL INITIAL (1970 DOLLARS)
4050 SOGICR     SYNTHETIC OIL AND GAS INVESTMENT COMPLETION RATE (1970 DOLLARS/YEAR)
4060 SOGIR      SYNTHETIC OIL AND GAS INVESTMENT RATE (1970 DOLLARS/YEAR)
4070 SOGPC      SYNTHETIC OIL AND GAS PRODUCTION CAPACITY (BTU/YEAR)
4080 SOGPR      SYNTHETIC OIL AND GAS PRODUCTION RATE (BTU/YEAR)
4090 SPC        SYNTHETICS PRODUCTION COST (1970 DOLLARS/BTU)
4100 SPD        SAFETY PERCEPTION DELAY (YEARS)
4110 SPDT       SAFETY PERCEPTION DELAY TIME (YEARS)
4120 SPMD       SURFACE PRODUCTION MULTIPLIER FROM DEPLETION (DIMENSIONLESS)
4130 SPMDT      SPMD TABLE
4140 SS         SAFETY STANDARD (FATALITIES/MILLION MAN-HOURS)
4150 SS69       VALUE OF SS AFTER TIME=1969 (FATALITIES/MILLION MAN-HOURS)
4160 SSH        SAFETY STANDARD HISTORICAL (FATALITIES/MILLION MAN-HOURS)
4170 SSN        NEW SAFETY STANDARD (FATALITIES/MILLION MAN-HOURS)
4180 SSP        VALUE OF SS AFTER TIME=PYEAR (FATALITIES/MILLION MAN-HOURS)
4190 SYNCCR     SYNTHETICS CAPACITY-CAPITAL RATIO (BTU/DOLLAR-YEAR)
4200 TEG        TOTAL ELECTRICITY GENERATION (BTU/YEAR)
4210 TEI        TOTAL ENERGY INVESTMENTS (DOLLARS/YEAR)
4220 TYEAR      YEAR OF TARIFF POLICY (YEAR)
4230 UC         UNDERGROUND CAPITAL (1970 DOLLARS)
4240 UC70       UNDERGROUND CAPITAL IN 1970 (1970 DOLLARS/YEAR)
4250 UCAF       UNDERGROUND COST ANNUALIZATION FACTOR (FRACTION/YEAR)
4260 UCC        UNDERGROUND COAL COST (1970 DOLLARS/BTU)
4270 UCCAF      UTILITIES CAPITAL COST ANNUALIZATION FACTOR (FRACTION/YEAR)
4280 UCDPLR     UNDERGROUND COAL DEPLETION RATE (BTU/YEAR)
4290 UCDR       UNDERGROUND CAPITAL DEPRECIATION RATE (1970 DOLLARS/YEAR)
4300 UCI        UNDERGROUND CAPITAL INITIAL (1970 DOLLARS)
4310 UCICR      UNDERGROUND CAPITAL INVESTMENT COMPLETION RATE (1970 DOLLARS/YEAR)
4320 UCIR       UNDERGROUND CAPITAL INVESTMENT RATE (1970 DOLLARS/YEAR)
4330 UCLS       UNDERGROUND COAL LABOR SUPPLY (MEN)
4340 UCLS70     UNDERGROUND COAL LABOR SUPPLY IN 1970 (MEN)
4350 UCLSI      UNDERGROUND COAL LABOR SUPPLY INITIAL (MEN)
4360 UCOC       UNDERGROUND CAPITAL AND OPERATING COST (1970 DOLLARS/BTU)
```

```
4370 UCPC      UNDERGROUND CAL PRODUCTION CAPACITY (BTU/YEAR)
4380 UCPC70    UNDERGROUND COAL PRODUCTION CAPACITY IN 1970 (BTU/YEAR)
4390 UCPR      UNDERGROUND COAL PRODUCTION RATE (BTU/YEAR)
4400 UCR       UNDERGROUND COAL RESOURCES (BTU)
4410 UCRI      UNDERGROUND COAL RESOURCES INITIAL (BTU)
4420 ULC       UNDERGROUND LABOR COST (1970 DOLLARS/3TU)
4430 UMCT      UNDERGROUND MINE CONSTRUCTION TIME (YEARS)
4440 UOGI      UNRESTRICTED OIL AND GAS IMPORTS (BTU/YEAR)
4450 UOGP      UNREGULATED OIL AND GAS PRICE (1970 DOLLARS/BTU)
4460 UPMD      UNDERGROUND PRODUCTION MULTIPLIER FROM DEPLETION (DIMENSIONLESS)
4470 UPMDT     UPMD TABLE
4480 UPRAT     UNREGULATED PRICE RATIO (DIMENSIONLESS)
4490 UPRATT    UPRAT TABLE
4500 UW        UNDERGROUND WAGES (1970 DOLLARS/MAN-YEAR)
4510 UWH       VALUE OF UW BEFORE TIME=PYEAR (1970 DOLLARS/MAN-YEAR)
4520 UWP       VALUE OF UW AFTER TIME=PYEAR (1970 DOLLARS/MAN-YEAR)
```

References

AEC 1972   Atomic Energy Commission. *Nuclear power 1973-2000* (WASH 1139(72)). Washington, D.C.: U.S. Government Printing Office.

AEC 1974   Atomic Energy Commission. *Nuclear power growth, 1974-2000* (WASH-1139). Washington, D.C.: U.S. Government Printing Office.

Alfven 1972   Alfven, Hannes. Energy and environment. *Bulletin of the Atomic Scientist*, 28, No. 5, p. 5, May 1972.

Basic Petroleum Data 1975   *Basic Petroleum Data Book.* American Petroleum Institute, 2101 L St. N.W., Washington, D.C. 20037.

Battelle 1975   Battelle Columbus Laboratories. *Status of stack gas technology for SO_2 control.* EPRI Report No. 209, Electric Power Research Institute, 3412 Hillview Ave., Palo Alto, Calif 94304.

Blueprint 1972   Blueprint for Survival. *The Ecologist*, Vol. 2, No. 1, January 1972.

Bureau of the Census 1972   Bureau of the Census. Illustrative population projections for the United States: the demographic effects of alternative paths to zero growth. In *Current population reports*, Series P-25, No. 480. Washington, D.C.: U.S. Government Printing Office.

BW 1975   Utilities: weak point in the energy future. *Business Week*, 20 Jan 1975.

CED 1974   Committee for Economic Development. *Achieving energy independence.* CED, 477 Madison Ave, New York, N.Y. 10022.

CEQ 1973   Council on Environmental Quality, Executive Office of the President. Prepared for U.S. Congress, Senate,

Committee on Interior and Insular Affairs, pursuant to Senate Resolution 45, 93rd. Congress, First Session, 1973, Serial No. 93-8 (92-43).

Chapman, Tyrrell, and Mount 1972
Chapman, Duane; Tyrrell, T.; and Mount, T. Electricity demand growth and the energy crisis. *Science* 178:703.

Chase 1974a
Chase Manhattan Bank. *Capital investments of the world petroleum industry, 1974.* Chase Manhattan Bank, 1 Chase Manhattan Plaza, New York, N.Y. 10015.

Chase 1974b
Chase Manhattan Bank. *Financial analysis of a group of petroleum companies*, by Dobies, R.S., Anderson, N.J., and Sparling, R.C. Energy Economics Division, Chase Manhattan Bank, 1 Chase Manhattan Plaza, New York, N.Y. 10015.

Cherniavsky 1974
Cherniavsky, E.A. *Energy systems analysis and technology assessment program.* Report No. BNL18184-R, Brookhaven National Laboratory, Upton, N.Y. 11973.

CIIA 1972
Committee on Interior and Insular Affairs, House of Representatives, U.S. Congress. *Energy 'demand' studies—an analysis and appraisal.* Washington, D.C., U.S. Government Printing Office.

CNR 1975
Air quality and stationary source emission control. A report by the Commission on Natural Resources, National Academy of Sciences, National Academy of Engineering, and National Research Council to the United States Senate Committee on Public Works. Washington, D.C.: U.S. Government Printing Office.

Coal Data 1974
Coal Data 1974 (and previous years: *Bituminous Coal Data*). National Coal Association, 1130 Seventeenth St. N.W., Washington, D.C. 20036.

Coal Facts 1974
Coal Facts 1974-1975. National Coal Association, 1130 Seventeenth St. N.W., Washington, D.C. 20036.

COMRATE 1975a
Committee on Mineral Resources and the Environment. *Mineral resources and the environment.* National Academy of Sciences, 2101 Constitution Ave. N.W., Washington, D.C. 20418.

COMRATE 1975b
Committee on Mineral Resources and the Environment. *Reserves and resources of uranium in the United States.* National Academy of Sciences, 2101 Constitution Ave. N.W., Washington, D.C. 20418.

Conoco 1973
Continental Oil Co. Coal's near-term outlook, by R.E. Bailey, in *Coal and the energy shortage.* Continental Oil Co., High Ridge Park, Stamford, Conn. 06904.

Cook 1974
Cook, Donald. As quoted in: Electric utilities face a price dilemma, *Business Week*, 2 Feb 1974.

Creamer, Dobrovolsky, and Borenstein 1960   Creamer, D., Dobrovolsky, S.P., and Borenstein, I. *Capital in manufacturing and mining.* Princeton, N.J.: Princeton University Press.

CRS 1973   Congressional Research Service. *Factors affecting the use of coal in present and future energy markets.* Prepared for U.S. Congress, Senate, Committee on Interior and Insular Affairs, pursuant to Senate Resolution 45, 93rd. Congress, 1st Session, 1973, Serial No. 93-9 (92-44).

CRS 1975   Congressional Research Service. *Factors affecting coal substitution for other fossil fuels in electric power production and industrial uses.* Prepared for United States Congress, Senate, Committee on Interior and Insular Affairs, pursuant to Senate Resolution 45, 94th Congress, 1st Session, 1975, Serial No. 94-17 (92-107).

Daly 1973   Daly, H.E., ed. *Toward a steady-state economy.* San Francisco: W.H. Freeman.

Day 1975   Day, M.C. Nuclear energy: a second round of questions. *Bulletin of the Atomic Scientists*, 31:52.

EEI 1974a   *Statistical yearbook of the electric utility industry for 1974* (and earlier years). Publication No. 75-39, Edison Electric Institute, 90 Park Ave., New York, N.Y. 10016.

EEI 1974b   *Report on equipment availability for the ten-year period 1964-1973.* Edison Electric Institute, 90 Park Ave., New York, N.Y. 10016.

EPA 1974   *National strategy for control of sulfur oxides from electric power plants.* Environmental Protection Agency, July 1974.

ERDA 1975   Energy Research and Development Administration. *A national plan for energy research, development, and demonstration: creating energy choices for the future* (ERDA-48). Washington, D.C.: U.S. Government Printing Office.

ERDA 1976   Energy Research and Development Administration. *A national plan for energy research, development, and demonstration: creating energy choices for the future* (ERDA 76-1). Washington, D.C.: U.S. Government Printing Office, Stock No. 052-010-00478-6.

Ervik 1974   Ervik, L.K. *The causal structure and implicit assumptions of alternative production functions.* DSD #19, System Dynamics Group, Thayer School of Engineering, Dartmouth College, Hanover, N.H. 03755.

Ervik 1975   Ervik, L.K. *Capital scarcity and the dynamics of growth in the United States coal industry.* DSD #22, System Dynamics Group, Thayer School of Engineering, Dartmouth College, Hanover, N.H. 03755.

FEA 1974   Federal Energy Administration. *Project indepen-dence*. Washington, D.C.: U.S. Government Printing Office, Stock No. 4118-00029.

FEA 1976   Federal Energy Administration. *1976 National energy outlook*. Washington, D.C.: U.S. Government Printing Office, Stock No. 041-018-00097-6.

FEA-CTF 1974   Federal Energy Administration—Coal Task Force. *Project independence blueprint, final task force re-port—coal*. Washington, D.C.: U.S. Government Print-ing Office.

FEA-NTF 1974   Federal Energy Administration—Nuclear Task Force. *Project independence blueprint, final task force re-port-nuclear energy*. Washington, D.C.: U.S. Govern-ment Printing Office.

FEA-STF 1974   Federal Energy Administration—Interagency Task Force on Synthetic Fuels from Coal. *Project indepen-dence blueprint—synthetic fuels from coal*. Washing-ton, D.C.: U.S. Government Printing Office, Stock No. 4118-00010.

Ford 1974a   Ford Energy Policy Project. *A time to choose: Ameri-ca's energy future*. Cambridge, Mass: Ballinger.

Ford 1974b   Ford Energy Policy Project. *Exploring energy choices*. The Energy Policy Project, P.O. Box 23212, Washington, D.C. 20024.

Ford, F.A. 1973a   Ford, F.A. *The dynamics of environmental standards: a case study of sulfur dioxide emissions*. DSD #10, System Dynamics Group, Thayer School of Engineer-ing, Dartmouth College, Hanover, N.H. 03755.

Ford, F.A. 1973b   Ford, F.A. *On the reduction of the peak demand for electricity in the State of Vermont*. DSD #43, Sys-tem Dynamics Group, Thayer School of Engineering, Dartmouth College, Hanover, N.H. 03755.

Ford, F.A. 1975   Ford, F.A. *A dynamic model of the United States electric utility industry, 1950-2010*. DSD #28, Sys-tem Dynamics Group, Thayer School of Engineering, Dartmouth College, Hanover, N.H. 03755.

Forrester 1961   Forrester, J.W. *Industrial dynamics*. Cambridge, Mass.: M.I.T. Press.

Forrester 1968   Forrester, J.W. *Principles of systems*. Cambridge, Mass.: Wright-Allen Press.

Forrester 1971   Forrester, J.W. Counterintuitive behavior of social systems. *Technology Review*, 73, No. 3, p. 53.

Forrester 1973   Forrester, J.W. *Confidence in models of social be-havior with emphasis on system dynamics models*. Memo #D-1967, M.I.T. System Dynamics Group, Massachusetts Institute of Technology, Cambridge, Mass.

Gas Facts 1974   *Gas Facts.* American Gas Association, Department of Statistics, 1515 Wilson Boulevard, Arlington, Va. 22209. Earlier years also.

Goldston 1973   Goldston, Eli. In Eastern Gas and Fuel Associates news release, 31 Oct 1973.

Gouse 1973   Gouse, S.W. Jr. *A program of research, development, and demonstration for enhancing coal utilization to meet national energy needs.* Environmental Studies Institute, Carnegie-Mellon University, Pittsburgh, Pa. 15213.

Hammond and Zimmerman 1975   Hammond, O. and Zimmerman, M.B. The economics of coal-based synthetic gas. *Technology Review*, 77:42.

Hass, Mitchell, and Stone 1974   Hass, J.E., Mitchell, E.J., and Stone, B.K. *Financing the energy industry.* Cambridge, Mass.: Ballinger.

Hendricks 1965   Hendricks, T.A. *Resources of oil, gas, and natural gas liquids in the United States and the world* (USGS Circular 522). Washington, D.C.: U.S. Government Printing Office.

Hubbert 1956   Hubbert, M.K. *Nuclear energy and the fossil fuels.* Publication No. 95, Shell Development Co., Exploration and Production Research Division, Houston, Texas.

Hubbert 1967   Hubbert, M.K. Degree of advancement of petroleum exploration in the United States. *Bulletin of the American Association of Petroleum Geologists*, 51:2207.

Hubbert 1969   Hubbert, M.K. Energy resources. In *Resources and man* by the Committee on Resources and Man, National Academy of Sciences—National Resource Council, San Francisco: W.H. Freeman.

Hubbert 1972   Hubbert, M.K., *United States energy resources, a review as of 1972.* Prepared for United States Congress, Senate, Committee on Interior and Insular Affairs, pursuant to Senate Resolution 45, 93rd Congress, 2nd. Session 1974, Serial No. 93-40 (92-75).

Joint Association Survey 1974   *Joint association survey of the United States oil and gas producing industry.* American Petroleum Institute, 1801 K St. N.W., Washington, D.C.

Joskow 1974   Joskow, Paul. *Inflation and environmental concern: structural change in the process of public utility price regulation.* Paper #128, M.I.T. Department of Economics, Massachusetts Institute of Technology, Cambridge, Mass. 02139.

Keystone 1974   *1974 Keystone Coal Industry Manual.* New York: McGraw-Hill.

Kneese 1973   Kneese, A.V. The Faustian bargain. *Resources*, No. 44. Resources for the Future, Inc., 1755 Massachusetts Avenue N.W., Washington, D.C. 20036.

Manne 1975   Manne, A.S. *United States options for a transition from oil and gas to synthetic fuels.* Discussion Paper #26D, Kennedy School of Government, Public Policy Program, Harvard University, Cambridge, Mass. 02139.

McKelvey 1973   McKelvey, V.E. Mineral resource estimates and public policy. In *United States mineral resources*, ed. D.A. Brobst and W.P. Pratt (USGS Professional Paper 820). Washington, D.C.: U.S. Government Printing Office, Stock No. 2401-00307.

Meadows, et al. 1972 Meadows, D.H., Meadows, D.L., Randers, J. and Behrens, W.W. III. *Limits to Growth.* New York: Universe Books.

Meadows et al. 1974 Meadows, D.L., Behrens, W.W. III, Meadows, D.H., Naill, R.F., Randers, J., Zahn, E.K. *Dynamics of growth in a finite world.* Cambridge, Mass.: Wright-Allen Press.

Minerals Yearbook *Minerals Yearbook* 1974. U.S. Bureau of Mines, U.S.
1974   Department of the Interior. Washington, D.C.: U.S. Government Printing Office.

MIT 1974   MIT Policy Study Group. Energy self-sufficiency: an economic evaluation. *Technology Review* 76:22.

MITRE 1975   MITRE. *An analysis of constraints on increased coal production.* No. MTR-6830, The MITRE Corp., McLean, Va. 22101.

Mobil 1974   Mobil. Oil and gas resources: did USGS gush too high? *Science* 185:127.

NAE 1974   National Academy of Engineering. *United States energy prospects: an engineering viewpoint.* National Academy of Sciences, 2101 Constitution Ave. N.W., Washington, D.C. 20418.

Naill 1974   Naill, R.F., Parameterizing social systems models. In *Methodological aspects of social system simulation*, D.L. Meadows, ed., DSD #14 System Dynamics Group, Thayer School of Engineering, Dartmouth College, Hanover, N.H. 03755.

Naill 1976   Naill, R.N. *COAL1: a dynamic model for the analysis of United States energy policy.* Doctoral thesis submitted to Thayer School of Engineering. DSD #54, System Dynamics Group, Thayer School of Engineering, Dartmouth College, Hanover, N.H. 03755.

Nathan 1975

Nathan, R.R. Testimony of Robert R. Nathan on petroleum pricing before the Senate Committee on the Interior. Robert R. Nathan Associates, 1200 Eighteenth St. N.W., Washington, D.C. 20036.

NPC 1972

United States energy outlook. National Petroleum Council, 1625 K. St. N.W., Washington, D.C. 20006.

NPC 1973

United States energy outlook: coal availability. National Petroleum Council, 1625 K. St. N.W., Washington, D.C. 20006.

NSC 1962

Accident Facts 1961. Chicago: National Safety Council.

Penn State 1973

Institute for Research on Human Resources. *The demand for and supply of manpower in the bituminous coal industry for the years 1985 and 2000.* Pennsylvania State University, University Park, Pa. 16802.

Rattein and Eaton 1976

Rattein, S. and Eaton, D. Oil shale: the prospects and problems of an emerging energy industry. *Annual Review of Technology* 1:153.

Ray 1973

Ray, Dixie Lee. *The nation's energy future* (WASH-1281). Washington, D.C.: U.S. Government Printing Office.

Risser 1973

Risser, H.E. The United States energy dilemma: the gap between today's requirements and tomorrow's potential. *Environmental Geology Notes No. 64*, Illinois State Geological Survey, Urbana, Ill. 61801.

Ross and Williams 1975

Ross, M.H. and Williams, R.H. *Assessing the potential for fuel conservation.* Institute for Public Policy Alternatives, State University of New York, Albany, N.Y. 12210.

Secrest and Burzlaff 1974

Secrest, Lee and Burzlaff, B. Financial dynamics of United States investor-owned electric utilities. In *Proceedings of the 1974 Summer Computer Simulation Conference*, Houston, Texas.

SITF 1975

Synfuels Interagency Task Force. *Recommendations for a synthetic fuels commercialization program.* A report submitted to the President's Energy Resources Council. Washington, D.C.: U.S. Government Printing Office, Stock No. 041-001-00111-3.

Spritzer 1972

Spritzer, R.S. Changing elements in the natural gas picture: implications for the regulatory scheme. In *Regulation of the natural gas producing industry*, ed. K.C. Brown. Baltimore: Johns Hopkins Press.

Steam-Electric Plant Factors 1975

Steam-Electric Plant Factors 1975 (and previous years). National Coal Association, 1130 Seventeenth St. N.W., Washington, D.C.

Teller 1975

Teller, Edward. *Energy: a plan for action.* Committee on Critical Choices for Americans, 22 W. 55th St., New York, N.Y. 10019.

Theobald, Schwein-
furth, and Duncan
1972

Theobald, P.K., Schweinfurth, S.P., and Duncan, D.C. *Energy resources of the United States* (USGS Circular 650). Washington, D.C.: U.S. Government Printing Office.

Train 1971

Train, Russel (Chairman, Council on Environmental Quality). Testimony on regulation of surface mining before Committee on Interior and Insular Affairs, House of Representatives, U.S. Congress. Washington, D.C.: U.S. Government Printing Office.

USBM 1971

U.S. Bureau of Mines. *Strippable reserves of bituminous coal and lignite in the United States* (Information Circular 8531). Washington, D.C.: U.S. Government Printing Office, Stock No. 2404-1020.

USBM 1972

U.S. Bureau of Mines. *Cost analyses of model mines for strip mining coal in the United States* (Information Circular 8535). Washington, D.C.: U.S. Government Printing Office, Stock No. 2404-1048.

USBM 1974

U.S. Bureau of Mines, Division of Fossil Fuels. Demonstrated coal reserve base of the United States on January 1, 1974. In *Mineral Industry Survey*, Washington, D.C.: Department of the Interior, Bureau of Mines.

USDI 1972

U.S. Department of the Interior. *United States energy through the year 2000* by Dupree, W.G., Jr. and West, J.A. Washington, D.C.: U.S. Government Printing Office, Stock No. 2400-00775.

USDI 1975

U.S. Department of the Interior. *United States energy through the year 2000 (Revised)* by Dupree, W.G., Jr. and Corsentino, J.S. Washington, D.C.: Department of the Interior, Bureau of Mines.

USGS 1967

U.S. Geological Survey. *Coal resources of the United States* by Paul Averitt (USGS Bulletin 1275). Washington, D.C.: Government Printing Office.

USGS 1973

U.S. Geological Survey. *United States mineral resources* by D.A. Brobst and W.P. Pratt (USGS Professional Paper 820). Washington, D.C.: U.S. Government Printing Office.

USGS 1975

Geological estimates of undiscovered recoverable oil and gas resources in the United States (USGS Circular 725). U.S. Geological Survey, National Center, Reston, Virginia, 22092.

Watson 1963

Watson, D.C. *Price theory and its uses.* New York: Houghton Mifflin Co.

Weinberg 1972

Weinberg, A.M. Social institutions and nuclear energy. *Science* 177:27.

Zapp 1961

Zapp, A.D. (unpublished), *World petroleum resources, domestic and world resources of fossil fuels, radioactive minerals, and geothermal energy.* Preliminary reports prepared by members of the U.S. Geological Survey for the Natural Resources Subcommittee of the Federal Council of Science and Technology, Nov. 28, 1961.

Zapp 1962

Zapp, A.D. *Future petroleum producing capacity of the United States* (USGS Bulletin 1142-H). Washington, D.C.: U.S. Government Printing Office.

Zimmerman 1975

Zimmerman, M.B. *Long-run mineral supply: the case of coal in the United States.* Ph.D. Thesis, Massachusetts Institute of Technology, Cambridge, Massachusetts.

Index

About the Author

Roger F. Naill is an Assistant Professor of Engineering at the Thayer School of Engineering, Dartmouth College.

He received an A.B. in Physics from Princeton University in 1969 and an M.S. degree in Management from the Alfred P. Sloan School of Management at M.I.T. He holds a Ph.D. in Engineering Sciences from Dartmouth College.

An associate engineer with the Raytheon Company from 1969 to 1971, he then joined the System Dynamics Group at M.I.T. which produced *The Limits to Growth* study. He has co-authored two subsequent volumes on growth issues: *Toward Global Equilibrium: Collected Papers* and *The Dynamics of Growth in a Finite World—A Technical Report of World3*. A founding member of the System Dynamics Group at Dartmouth College, he is currently engaged in teaching and modeling energy and economic systems.